# SRA
# Essentials for
# Algebra

*A Direct Instruction Approach*

## Teacher's Guide

Siegfried Engelmann
Bernadette Kelly
Owen Engelmann

Mc Graw Hill **SRA**

*Columbus, OH*

## Acknowledgments

The authors are grateful to the following people for their input in the field-testing and preparation of *SRA Essentials for Algebra: A Direct Instruction Approach:*

Allison Bolen
Crystal Chambers
Ann Desjardins
Laura Doherty
Karen Fierman
Vicky Jennings
Cheryl Krueger
Katie Krueger
Jeff Lalor
Rich Lalor
Karen Lancaster
Patrice Larson
Patti Lingerfelt
Kathryn McDonald
Denny McKee
Billie Overholser
Christopher Perdue
Pat Pielaet
Mary Rosenbaum
Pat Smith
Tina Wells

**Front Cover Photo:** © Lester Lefkowitz/Getty Images, Inc.

## SRAonline.com

 **SRA**

Send all inquiries to this address:
SRA/McGraw-Hill
4400 Easton Commons
Columbus, OH 43219

ISBN: 978-0-07-602196-3
MHID: 0-07-602196-3

3 4 5 6 7 8 9 QPD 13 12 11 10 09 08

# Table of CONTENTS

# Program Summary

## Facts about *Essentials for Algebra*

| | |
|---|---|
| **Students who are appropriately placed in *Essentials for Algebra*** | Students who have passed the placement test |
| **Format of lessons** | Scripted presentations for all activites<br>Program designed for presentation to entire class |
| **Number of lessons** | 118 plus 2 test-preparation lessons |
| **Scheduled time for math period** | As a middle-school or junior-high-school program:<br>    55 minutes per period for teacher-directed activities<br>    Additional 45 minutes to 1 hour of independent work<br>      (homework)<br>As a high-school intervention program:<br>    90 minutes per period for teacher-directed activities<br>      and independent practice<br>    Additional 10 to 20 minutes of independent work<br>      (homework) |
| **Weekly schedule** | 5 periods |
| **Teacher material** | *Teacher's Guide*<br>*Presentation Book 1* (Lessons 1–60)<br>*Presentation Book 2* (Lesson 61–end of program)<br>*Answer Key*<br>*Board Display CD* |
| **Student material** | Program material:<br>    Workbook<br>    Textbook   &#124;  Additional materials:<br>                            calculator with $\pi$<br>                            function<br>                            ruler |
| **In-program tests** | Mastery Tests 1–11 (A Mastery Test follows lessons 6 and 15 and every tenth lesson thereafter, that is, lessons 25, 35, 45, and so on through lesson 115.)<br>Practice Tests 1 and 2 follow lesson 120. |

# Scope and Sequence for
# *Essentials for Algebra*

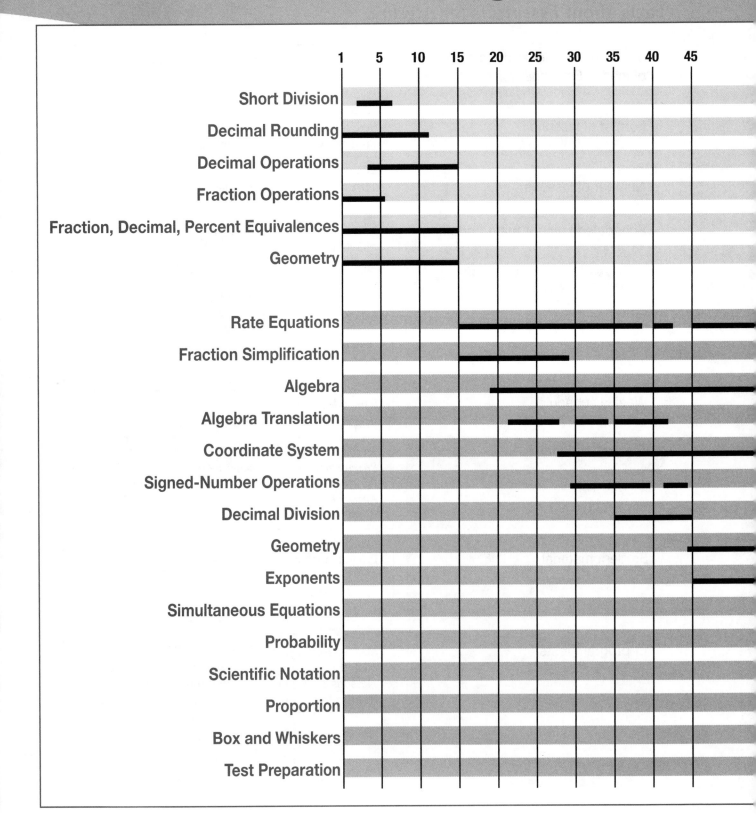

|  | 1 | 5 | 10 | 15 | 20 | 25 | 30 | 35 | 40 | 45 |
|---|---|---|---|---|---|---|---|---|---|---|
| Short Division | | | | | | | | | | |
| Decimal Rounding | | | | | | | | | | |
| Decimal Operations | | | | | | | | | | |
| Fraction Operations | | | | | | | | | | |
| Fraction, Decimal, Percent Equivalences | | | | | | | | | | |
| Geometry | | | | | | | | | | |
| Rate Equations | | | | | | | | | | |
| Fraction Simplification | | | | | | | | | | |
| Algebra | | | | | | | | | | |
| Algebra Translation | | | | | | | | | | |
| Coordinate System | | | | | | | | | | |
| Signed-Number Operations | | | | | | | | | | |
| Decimal Division | | | | | | | | | | |
| Geometry | | | | | | | | | | |
| Exponents | | | | | | | | | | |
| Simultaneous Equations | | | | | | | | | | |
| Probability | | | | | | | | | | |
| Scientific Notation | | | | | | | | | | |
| Proportion | | | | | | | | | | |
| Box and Whiskers | | | | | | | | | | |
| Test Preparation | | | | | | | | | | |

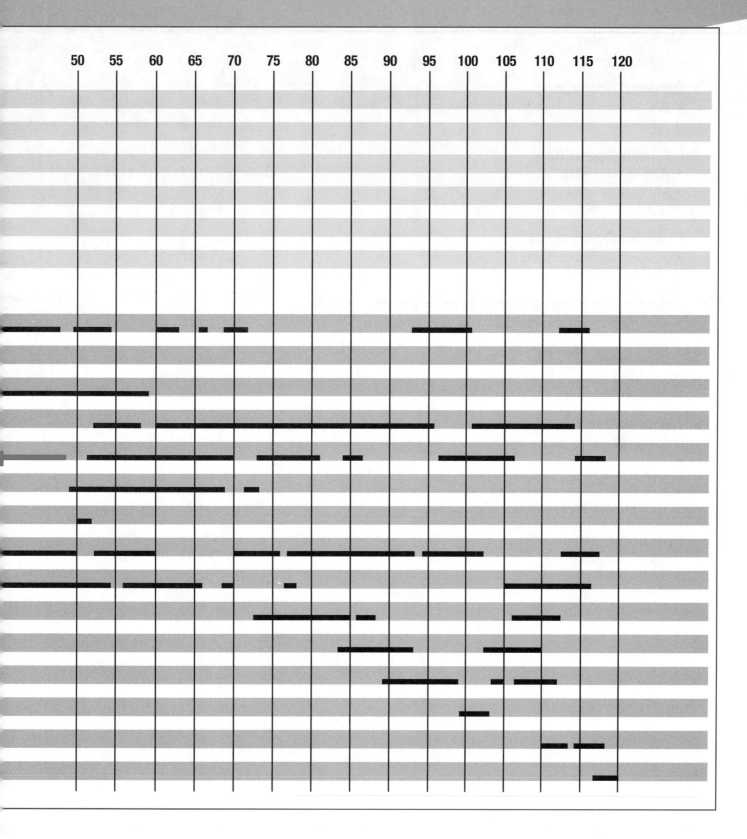

# Objectives Taught in
# *Essentials for Algebra*

| | Assessed in | Mastered by |
|---|---|---|
| **Short Division** | | |
| • Divide a 3-digit or 4-digit value by a single-digit value. | Test 1A | Lesson 7 |
| **Decimal Rounding** | | |
| • Round a decimal value to the nearest whole number, tenth, hundredth, or thousandth. | Test 1B | Lesson 16 |
| **Decimal Operations** | | |
| • Add or subtract decimal values. | Test 1A | Lesson 7 |
| • Multiply decimal values. | Test 1B | Lesson 16 |
| • Divide decimal values. | Test 4 | Lesson 46 |
| **Fraction Operations** | | |
| • Add or subtract fractions with like denominators. | Test 1A | Lesson 7 |
| • Multiply fractions. | Test 1A | Lesson 7 |
| **Fraction, Decimal, Percent Equivalences** | | |
| • Complete an equation to show equivalent fractions. | Test 1B | Lesson 16 |
| • Complete an equation to show a fraction and the equivalent mixed number. | Test 1A | Lesson 7 |
| • Complete a table to show a hundredths fraction and the equivalent decimal and percent values. | Test 1B | Lesson 16 |
| **Abbreviations** | | |
| • Write abbreviations for common standard and metric units. | Test 1A | Lesson 7 |
| **Problem Solving: Add/Subtract** | | |
| • Find the total cost of a purchase or the change received. | Test 1B | Lesson 16 |
| • Find the difference between two values. | Test 7 | Lesson 76 |
| **Fraction Simplification** | | |
| • Apply divisibility rules for 2, 3, 5, and 10. | Test 2 | Lesson 26 |
| • Simplify a fraction. | Test 2 | Lesson 26 |
| • Simplify fractions that are added or multiplied. | Test 3 | Lesson 36 |
| • Simplify a fraction with decimal values. | Test 5 | Lesson 56 |

| | Assessed in | Mastered by |
|---|---|---|

## Problem Solving: Rate Equations

| | Assessed in | Mastered by |
|---|---|---|
| • Write a letter equation from a question. | Test 2 | Lesson 26 |
| • Solve a complete problem. | Test 3 | Lesson 36 |
| • Solve a ratio problem. | Test 4 | Lesson 46 |
| • Solve a mixed set of rate-equation and classification problems. | Test 4 | Lesson 46 |
| • Solve a mixed set of rate-equation, classification, and comparison problems. | Test 5 | Lesson 56 |
| • Solve a problem that asks about the rate unit. | Test 6 | Lesson 66 |
| • Work a mixed set of problems, some of which ask about the rate unit. | Test 7 | Lesson 76 |
| • Use survey and sample data to estimate expected outcomes. | Test 10 | Lesson 106 |
| • Convert related units. | Test 11 | Lesson 116 |
| • Solve a rate-equation problem that involves unit conversion. | —— | Lesson 117 |

## Algebra

| | Assessed in | Mastered by |
|---|---|---|
| • Solve a missing-factor problem. | Test 2 | Lesson 26 |
| • Solve a one-step add/subtract problem. | Test 2 | Lesson 26 |
| • Solve a problem that involves multiplication by a reciprocal. | Test 3 | Lesson 36 |
| • Solve a two-step problem. | Test 3 | Lesson 36 |
| • Solve a problem that involves substitution. | Test 4 | Lesson 46 |
| • Solve a problem that adds or subtracts a whole number and a fraction. | Test 5 | Lesson 56 |
| • Solve a problem that involves like terms. | Test 5 | Lesson 56 |
| • Simplify an expression that involves the distribution of a number. | Test 6 | Lesson 66 |
| • Solve a problem that involves a negative letter term (multiply by $-1$). | Test 7 | Lesson 76 |
| • Solve a problem that involves two substitutions. | Test 7 | Lesson 76 |
| • Solve a problem that involves the distribution of a number. | Test 7 | Lesson 76 |
| • Solve a problem that involves the distribution of a letter. | Test 7 | Lesson 76 |
| • Solve a problem with letter terms on both sides of the equation. | Test 7 | Lesson 76 |
| • Solve a one- or two-step inequality ($>$, $<$). | Test 8 | Lesson 86 |
| • Solve a mixed set of equations and inequalities that involve substitution. | Test 8 | Lesson 86 |
| • Solve an equation with the unknown in the denominator of a fraction. | Test 8 | Lesson 86 |
| • Solve a pair of simultaneous equations by multiplying and combining equations. | Test 8 | Lesson 86 |
| • Solve a pair of simultaneous equations by substitution. | Test 9 | Lesson 96 |
| • Solve a pair of simultaneous equations for $x$ and $y$. Plot the lines and show the intersection. | Test 11 | Lesson 116 |

## Problem Solving: Algebra Translation

| | Assessed in | Mastered by |
|---|---|---|
| • Solve a classification problem (+ or −). | Test 3 | Lesson 36 |
| • Translate a comparison sentence into a letter equation (+, −, ×). | Test 3 | Lesson 36 |
| • Solve a comparison problem (+, −, ×). | Test 4 | Lesson 46 |
| • Solve a multiplication comparison problem involving a percent value. | Test 6 | Lesson 66 |
| • Solve a problem that asks about a fraction or percent of a group. | Test 7 | Lesson 76 |
| • Solve a classification problem that involves two multiplication equations. | Test 7 | Lesson 76 |
| • Translate a sentence into a letter or equation that involves two operations. | Test 7 | Lesson 76 |
| • Solve a problem that yields an equation involving two or more operations. | Test 8 | Lesson 86 |
| • Solve a problem that yields an inequality statement involving two or more operations. | Test 9 | Lesson 96 |
| • Translate a sentence that yields a combination sign ($\geq$, $\leq$). | Test 9 | Lesson 96 |
| • Solve a problem that yields a combination sign. | Test 9 | Lesson 96 |
| • Solve a mixed set of problems ($>$, $<$, $=$, $\leq$, $\geq$). | Test 10 | Lesson 106 |
| • Solve a two-step problem that asks about a fraction or percent of a group. | Test 10 | Lesson 106 |
| • Solve a problem that involves a percent increase or decrease. | Test 11 | Lesson 116 |
| • Solve a problem that generates a pair of simultaneous equations. | Test 11 | Lesson 116 |

## Coordinate System

| | Assessed in | Mastered by |
|---|---|---|
| • Plot a point from a description ($x = \blacksquare$, $y = \blacksquare$). | Test 3 | Lesson 36 |
| • Write an $x$ and $y$ equation for a point. | Test 3 | Lesson 36 |
| • Plot a point from coordinates ($\blacksquare$, $\blacksquare$). | Test 4 | Lesson 46 |
| • Write coordinates for a point (4 quadrants). | Test 4 | Lesson 46 |
| • Answer questions based on rate for lines on the coordinate system. | Test 11 | Lesson 116 |

## Signed-Number Operations

| | Assessed in | Mastered by |
|---|---|---|
| • Combine 2 values. | —— | Lesson 37 |
| • Combine more than 2 values. | Test 4 | Lesson 46 |
| • Multiply 2 values. | Test 5 | Lesson 56 |
| • Divide 2 values. | Test 6 | Lesson 66 |
| • Multiply and combine a string of values. | Test 6 | Lesson 66 |
| • Multiply more than 2 values. | Test 7 | Lesson 76 |

## Straight-Line Equations

| | Assessed in | Mastered by |
|---|---|---|
| • Complete an add/subtract function table and draw the line. | Test 4 | Lesson 46 |
| • Figure out the correct function for a table $(+, -, \times)$. | Test 4 | Lesson 46 |
| • Figure out the correct function for a table based on $x\left(\frac{y}{x}\right) = y$. | Test 5 | Lesson 56 |
| • Complete a multiplication function table with missing $x$ or $y$ values and draw the line. | Test 6 | Lesson 66 |
| • Solve a linear equation for $y$, and write an equation for the slope $(m = \blacksquare)$. | Test 6 | Lesson 66 |
| • Write the slope-intercept equation for a line shown on the coordinate system (+ slope, through zero). | Test 6 | Lesson 66 |
| • Write the slope-intercept equation for a line (+ slope, $+/-$ intercept). | Test 7 | Lesson 76 |
| • Write the slope-intercept equation for a line ($+/-$ slope, $+/-$ intercept). | Test 8 | Lesson 86 |
| • Plot a line on the coordinate system for a given equation ($+/-$ slope, $+/-$ intercept). | Test 8 | Lesson 86 |
| • Plot a line on the coordinate system that involves a whole-number slope. | Test 9 | Lesson 96 |
| • Substitute a given $x$ or $y$ value in a slope-intercept equation to figure out the corresponding $y$ or $x$ coordinate and plot the line. | Test 11 | Lesson 116 |
| • Identify the $y$ intercept for a line on the coordinate system. | Test 11 | Lesson 116 |

## Exponents

| | Assessed in | Mastered by |
|---|---|---|
| • Write the base and exponent for repeated multiplication. | Test 5 | Lesson 56 |
| • Write repeated multiplication for a base and exponent. | Test 5 | Lesson 56 |
| • Figure out the value for a base and exponent. | Test 5 | Lesson 56 |
| • Write the base and exponent for groups of repeated multiplication. | Test 5 | Lesson 56 |
| • Write the base and exponent for fractions showing repeated multiplication. | Test 5 | Lesson 56 |
| • Express a fraction involving repeated multiplication as a base with a positive or with a negative exponent. | Test 6 | Lesson 66 |
| • Rewrite a fraction with multiplied bases and exponents to show an equivalent fraction with all positive exponents. | Test 6 | Lesson 66 |
| • Simplify an expression by combining exponents. | Test 7 | Lesson 76 |
| • Rewrite an expression to show the value of a numerical base and exponent. | Test 7 | Lesson 76 |
| • Simplify a fraction that has a positive and negative exponent in both the numerator and denominator. | Test 8 | Lesson 86 |
| • Combine like terms that have exponents. | Test 10 | Lesson 106 |
| • Simplify an expression by multiplying and combining terms with exponents. | Test 11 | Lesson 116 |
| • Combine exponents and figure out the value of an expression involving a negative base. | Test 11 | Lesson 116 |

|  | Assessed in | Mastered by |
|---|---|---|

## Geometry
- Find the perimeter of a polygon.
- Find the area of a rectangle or triangle.
- Find the circumference or diameter of a circle.
- Find the radius of a circle, given the diameter.
- Find the area of a circle.
- Know the degrees in a circle, a right angle, and a straight line.
- Find the surface area of a box.
- Find the area of a trapezoid.
- Find the area of complex shapes.
- Find the volume of a rectangular prism.
- Find the volume of complex figures.

| | Assessed in | Mastered by |
|---|---|---|
| Find the perimeter of a polygon. | Test 1A | Lesson 7 |
| Find the area of a rectangle or triangle. | Test 1B | Lesson 16 |
| Find the circumference or diameter of a circle. | Test 5 | Lesson 56 |
| Find the radius of a circle, given the diameter. | Test 5 | Lesson 56 |
| Find the area of a circle. | Test 6 | Lesson 66 |
| Know the degrees in a circle, a right angle, and a straight line. | Test 7 | Lesson 76 |
| Find the surface area of a box. | Test 11 | Lesson 116 |
| Find the area of a trapezoid. | —— | Lesson 117 |
| Find the area of complex shapes. | —— | Lesson 118 |
| Find the volume of a rectangular prism. | —— | Lesson 118 |
| Find the volume of complex figures. | —— | Lesson 119 |

## Pythagorean Theorem

| | Assessed in | Mastered by |
|---|---|---|
| Identify the square root of a number or the whole numbers a square root lies between. | Test 8 | Lesson 86 |
| Complete an equation to show the square root or square of a number. | Test 8 | Lesson 86 |
| Find the missing side in a right triangle. | Test 9 | Lesson 96 |
| Solve word problems that involve distance and direction, some of which generate a right-triangle diagram. | Test 10 | Lesson 106 |

## Similar Triangles

| | Assessed in | Mastered by |
|---|---|---|
| Figure out a missing angle to determine whether or not two triangles are similar. | Test 9 | Lesson 96 |
| Figure out a corresponding side in a pair of similar triangles. | Test 9 | Lesson 96 |
| Figure out a corresponding side for right triangles shown on parallel lines. | Test 10 | Lesson 106 |
| Figure out a corresponding side in a pair of nested similar triangles. | Test 10 | Lesson 106 |
| Solve a word problem that generates a similar-triangle diagram. | Test 10 | Lesson 106 |

## Probability

| | Assessed in | Mastered by |
|---|---|---|
| Write a probability fraction for an event involving a spinner. | Test 9 | Lesson 96 |
| Solve a probability problem that asks about trials. | Test 9 | Lesson 96 |
| Solve a probability problem that asks about the object. | Test 9 | Lesson 96 |
| Compute the probability of independent events. | —— | Lesson 107 |
| Compute the probability of dependent events. | Test 11 | Lesson 116 |

## Scientific Notation

| | Assessed in | Mastered by |
|---|---|---|
| Write the scientific notation for a number (+ exponent). | Test 9 | Lesson 96 |
| Write a number from scientific notation (+ exponent). | Test 10 | Lesson 106 |
| Write a number from scientific notation (+/− exponent). | Test 11 | Lesson 116 |
| Write the scientific notation for a number (+/− exponent). | Test 11 | Lesson 116 |

## Proportion

| | Assessed in | Mastered by |
|---|---|---|
| Use a scale diagram to figure out the actual dimension of an object. | Test 10 | Lesson 106 |

## Box and Whiskers

| | Assessed in | Mastered by |
|---|---|---|
| Find the mean and median score for a population of scores. | Test 11 | Lesson 116 |
| Construct a box-and-whiskers plot for a population of scores. | —— | Lesson 119 |

# Overview

## For Whom Was *Essentials for Algebra* Developed?

*Essentials for Algebra* is designed for students in middle school or high school who are at risk of failing to meet graduation requirements in math. The program teaches essential pre-algebra content and provides students with an introduction to traditional Algebra I content that is cohesive and clear. In many districts, graduation requirements (as expressed either by standards or by the content of exit examinations) assume that students have learned basic geometry (Pythagorean theory, similar triangles), introductory algebra (possibly involving simultaneous equations), solution strategies involving straight-line equations on the coordinate system, exponents, signed numbers, facility with problems that refer to fractions, decimals and percents, data tables or graphs, and a wide range of word problems that involve rate, proportion, probability, and algebraic solutions.

This guide describes how *Essentials for Algebra* thoroughly addresses each of these topics.

## The Teaching Challenge Addressed by *Essentials for Algebra*

*Essentials for Algebra* attempts to provide quite a bit more than one year's instruction in a sequence that may be presented and taught to mastery in one year. Understand, however, that the typical student who qualifies for the program is greatly behind in understanding math. The typical portrait is a student who is completely confused about fractions and anything associated with fractions—who does not understand the relationship between fractions and whole numbers, who is not clear on the operations of combining, multiplying, or simplifying fractions, and who does not understand the relationship between fractions and division. These students perform at possibly the third- or fourth-grade level in math operations and understanding but are expected to perform in a pre-algebra or beginning algebra course.

Furthermore, they have a history of failure, so they tend not to like math, and they are not optimistic about their chances of learning what the teacher teaches.

The goal of *Essentials for Algebra* is to teach a solid foundation for a traditional Algebra I course and to teach students enough about algebra and other topics typically presented in high-school math exams that are based on graduation requirements. This is provided in a one-year course.

Obviously, the architecture of a program that is capable of achieving the goal of teaching much more than a year's worth of skills to lower performers in only one year will be quite different from that of a traditional developmental sequence. The architecture must make it possible to cut through detail, simplify, and sequence what is taught so that it begins on a level the students understand and progresses at a rate that permits them to learn. Certainly the sequence that achieves this goal may have weak spots, but it will also have inventions that not only save time but that permit students to learn and understand material they often do not learn in traditional sequences.

Perhaps the greatest departure that *Essentials for Algebra* makes from traditional sequences is that it provides explicit instruction for some skills and areas that traditionally receive only scant instruction. For instance, traditional sequences typically provide students with only spotty information about how to translate word problems into algebraic equations. Possibly, the information the students receive comes after they have struggled to work a problem and failed. The teacher then shows the solution and tries to answer questions. Students often do not receive practice on many examples of the same type, but next work problems that require different translations and different problem-solving steps. Some students learn from after-the-fact information; most low performers don't. *Essentials for Algebra* provides students with information about how to examine a problem, refer to specific details, and use those details to construct an equation (or equations) that will lead to a proper solution.

Understand that the program cannot perform miracles, but it can teach very low performers the pre-algebra content they have failed to learn in the past, and they can learn enough to perform respectably on math exams that measure basic algebra.

## Program Duration

The program contains 120 lessons and 11 in-program mastery tests. The ideal goal is to teach one lesson each period. If students are not firm on content that is being introduced, you will need to repeat parts of lessons or entire lessons. Also, in the later parts of the sequence, lessons are longer and tend to introduce content at a faster fate. For lower-performing students, teaching the program requires one full year and possibly more if many lessons or exercises must be repeated to achieve student mastery. Even if students are higher performers and start at lesson 16 rather than lesson 1, teaching the content to mastery may take the entire school year.

## Scheduling

The purpose for using the program implies when it is best taught. If *Essentials* is being offered for pre-algebra credit, it can be presented in the standard pre-algebra sequence. If *Essentials* is preparing students to meet graduation requirements that are tested in the tenth grade, the program should be taught in the ninth grade. If the students are particularly low (as indicated by their placement-test performance), a good plan would be to teach as much of the program as possible during the ninth grade and complete or review the program in the tenth grade. A fairly safe estimate for middle-school or junior-high groups is that with 155 school days and daily 55-minute periods, you should be able to present the program to mastery and present the accompanying test-preparation material. For high-school students, a daily double period is recommended.

*Essentials* is not designed as a supplemental program and should not be used as one. Some strategies taught in *Essentials* are incompatible with those taught in traditional pre-algebra programs.

# How the Program Is Different

*Essentials for Algebra* differs from traditional approaches in the process of program development and field testing, in its instructional design and scripted presentations, and in the language it uses to communicate with students.

## Field Tested

*Essentials* has been shaped through extensive field testing and revision based on difficulties students and teachers encountered. The field-test philosophy of *Essentials* is that, if teachers or students have trouble with material presented, the program is at fault. Revisions are made to alleviate the problems.

The field-test results of *Essentials for Algebra* disclose that if the teacher implements the program according to the presentation detail provided for each exercise, failed students will learn the content and become sufficiently proficient in the content to advance to the next levels of math instruction.

## Organization and Instructional Design

The organization of how skills are introduced, developed, and reviewed is based on the principles of Direct Instruction, which have been successful in various academic areas.

This program represents a sharp departure from the idiom of how to teach math through "discovery" or even through traditional programs that have a progression of "units" that are augmented by some form of cumulative review. These programs tend to assume that by focusing several *lessons* on a single topic, such as translating algebra word problems, students receive the information and amount of practice they need to learn to perform the translation.

This design does not provide enough practice and is not well configured to hold students to a high standard of mastery. In fact, mastery is not a serious issue with this design because the next unit typically has nothing to do with the previous one.

*Essentials for Algebra* does not follow this format for the following reasons:

a) During a period, it is not productive to work only on a single topic. If a lot of new information is being presented, students may become overwhelmed. A better procedure, one that has been demonstrated to be superior in studies of learning and memory, is to distribute the practice, so instead of working for 55 minutes on a single topic, students work each day for possibly 10 minutes on each of four or five topics.

b) When full-period topics are presented, it becomes very difficult for a teacher to provide sufficient practice and review on the latest skills that have been taught. If these skills are not used and reviewed, students' performance will deteriorate, and the skills will have to be retaught when they reappear. A more productive organization is to present work on skills continuously (not discontinuously) so that students work on a particular topic (such as rate and proportion problems) for part of 40 lessons, not for five or six entire lessons at a time. In the context of continuous development of skills, review becomes automatic and reteaching becomes unnecessary because students use the skills in every lesson.

c) If teaching occurs in only several lessons, it is difficult for students to become "automatic" in performing operations. A better method is to develop skills and concepts in small steps so that students are not required to learn as much new material at a time and so they receive a sufficient amount of practice to become facile or automatic in applying it.

d) When skills are not developed continuously, students and teachers may develop negative attitudes. Students may think they are not expected to "learn" the new material. Teachers often become frustrated because they understand that students need much more practice, but they are unable to provide it if they are to move through the program at a reasonable rate. Continuous development of skills solves this problem because students learn very quickly that what is presented is used in this lesson, the next lesson, and many subsequent lessons.

In *Essentials for Algebra*, skills are organized in **tracks**. A track is an ongoing development of a particular topic. Within each lesson, work from four to six tracks is presented. The teaching presentations are designed with the goal of presenting the entire lesson in 55–60 minutes (although some lessons may run longer, and a double period is recommended for lower performers).

From lesson to lesson, the work on new skills develops a small step at a time so that students are not overwhelmed with new information and receive enough practice both to master skills and to become facile with them. Only about 15% of what appears in a typical lesson of *Essentials* is new. The rest is either work on problem types that have been introduced in the preceding lessons or slight expansions or new applications that build on what was taught earlier. For instance, in lesson 50 (see pages 29 to 38 of this guide), the program presents problems from five different tracks: *exponents, rate equations, signed-number multiplication, geometry,* and *algebra.* In the preceding lesson and the one that follows, students work on the same five tracks. In lesson 53, an additional track, *function tables,* is added. A particular track sequence may continue for several lessons, be reviewed in the independent work for a number of lessons, and then return to the structured part of the lesson for continued expansion. The scope and sequence chart on pages 2 and 3 shows the tracks that occur in the 120-lesson sequence.

A design feature of the instruction in all the tracks is to teach in a way that requires relatively less learning. The program addresses all the types, including a wide range of word problems; however, the procedure for working problems of each type is as uniform and simple as the authors know how to make it.

The Tracks section of this guide (page 49) presents an overview of the design strategies used in each track to minimize the amount of new material that is taught. The strategies taught in the *rate-equation* track illustrate the savings that are achieved by uniform strategies. The procedures students learn are fairly simple but allow them to process a full range of proportion, rate, and unit-conversion problems. Here's a simple rate problem:

> **A train travels at an average rate of 46 miles per hour. The train travels 218 miles. How long does that trip take?**

Students work this problem by constructing a rate equation that is *based on the question.* The question

asks about distance. The unit is hours. So students construct a simple equation for hours:

$$h = h$$

They complete a valid rate equation by showing the other unit, miles:

$$m \left( \frac{h}{m} \right) = h$$

Note that students construct a non-conventional equation. They do not have to memorize equations such as "distance equals rate times time." They simply observe that the problem refers to the units that are related (46 miles in each hour) and construct the equation on the basis of the unit named in the question.

This approach applies to a range of problems that ask about ratios, proportions, unit conversions, and rate. Because all these subtypes admit to the same solution strategy, the strategy is taught once and then is systematically applied to various subtypes.

During a lesson in which this track appears, students are able to work four or five word problems of the same type in possibly 12 to 15 minutes. After days of this practice, they have a fairly solid understanding of how to approach the problem type, how to construct a workable equation, and how to answer the question the problem asks.

Because the program is designed to teach all the discriminations students need to work problems, the track that teaches the solution strategy does not begin with word problems. It begins with a basic principle of multiplication, which is that $X$ times some value equals $Y$. That value is expressed as $Y$ over $X$.

$$x \left( \frac{y}{x} \right) = y$$

Students practice working on problems of this basic type with numbers and letters:

$$5 \, (\blacksquare) = 6$$

$$3 \, (\blacksquare) = r$$

Students also practice generating a multiplication equation from a fraction. Given 4/9, they write this equation:

$$9 \left( \frac{4}{9} \right) = 4$$

They also work problem sets in which they discriminate between problems like these:

$$6 \, (p) = 7 \text{ and } 4 \left( \frac{6}{7} \right) = p$$

(Students solve for $p$.)

Next, students learn to construct letter equations from questions, not entire problems, just questions.

**How many miles does she jog in 8 days?**

**22 bags weigh how many pounds?**

Each question names two units. Students construct the letter equation based on the unit name that answers the question:

$$d \left( \frac{m}{d} \right) = m \qquad b \left( \frac{p}{b} \right) = p$$

Students are able to construct several equations in a short period of time (5 equations in 5 minutes or less).

Before complete word problems are introduced, students learn to identify the related units the problem gives (the two values that go in the fraction).

The final step is for students to apply the procedures to word problems that present a full range of wording variations. Further details of the track are provided on pages 63–77.

The point emphasized here is that all the component skills students need to work complicated problems are introduced and sequenced in a manner that maximizes practice for each component of the strategy.

**Note:** *Essentials for Algebra* is designed to teach all the skills and discriminations that are introduced; however, the material must be presented as specified, in sequence, with all parts of all lessons presented in the specified lesson order. The reasons for these conventions follow.

The authors developed this program from the premise that failed students and those who are at risk of failure are capable of learning practical math and basic algebraic operations if the program provides the following:

a) articulate information about how to discriminate one problem type from others that are similar;

b) far more practice on each problem type than traditional programs provide;

c) a progression of introducing and expanding problem types that do not overwhelm the student with new information in any lesson;

d) techniques that maximize the amount of time that students are engaged in practice (not listening to the teacher or puzzling over how to solve a problem).

One result of these design features is that students learn the detailed anatomy of different problem types. Because this information is provided before students work a particular problem type, the teacher is provided with all the information needed to correct errors quickly. The teacher should not have to introduce rules, concepts, or technical details to correct errors because the essential ones have been presented earlier. To correct, the teacher points out where the students' performance departed from the model they were to follow.

## SOLVING PROBLEMS A STEP AT A TIME

Another design feature of *Essentials for Algebra* is that it teaches all the necessary procedures for solving problems. The first part of the program may give both teacher and students the impression that there are "busy-work" steps, but all the procedures taught earlier will be used later.

Students don't take shortcuts. They write and rewrite problems according to the rules and the directions the teacher provides.

Students who learn to follow the solution steps specified for the earlier problems develop various habits that make the later problems far less difficult than they are for students who do not have a repertoire of problem-solving strategies that involve writing and rewriting equations.

The program provides students with solution strategies that permit them to be flexible. Students who complete the program will know three or more strategies for writing equations that are strictly based on the wording of a problem. Students who have an understanding of these strategies have the potential to tackle problems that don't appear in the program by extending and applying what they know. They have the tools needed to experiment and the ability to see the extent to which approaches lead them closer to a solution to the problem. Students who do not have knowledge of how to decipher problems are preempted from systematically constructing and experimenting with different equations. (See the Algebra Translation track on page 89 for examples of constructing letter equations.)

## Scripted Presentations

The format of this program is foreign to most middle- and high-school teachers. Each lesson is scripted. The script indicates the exact wording the teacher is to use in presenting the material and correcting student errors. Certainly, you will deviate some from the exact wording; however, until you know why things are phrased as they are, you should follow the exact wording.

In *Essentials for Algebra*, you first present material in a structured sequence that requires students to respond verbally. This technique permits you to present tasks to your students at a higher rate than is possible if students write the problems. Also, it provides you with information about which students are responding correctly and which need more repetition.

After students respond to a series of verbal tasks, you present written work.

Part of an exercise follows. Students have just written the signs in the answer for six problems that combine signed numbers. (They did not work the problems; they simply wrote the sign in the answers.) Now you present a series of questions that address the number part of the answer.

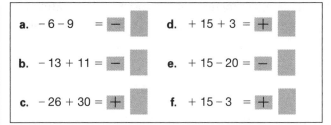

- Here's the rule for the number part: If the signs are the same, you add.
- Listen again: If the signs are the same, you add.
- Say that rule. (Signal.) *If the signs are the same, you add.*

d. If the signs are different, you subtract.
- Say that rule. (Signal.) *If the signs are different, you subtract.*

e. Problem A: − 6 − 9.
- Are the signs of 6 and 9 the same or different? (Signal.) *The same.*
- The signs are the same, so what do you do? (Signal.) *Add.*
- (Repeat step e until firm.)

f. Yes, 6 + 9 gives the number part of the answer.

g. Problem B: − 13 + 11.
- Are the signs of 13 and 11 the same or different? (Signal.) *Different.*
- The signs are different, so what do you do? (Signal.) *Subtract.*
- (Repeat step g until firm.)

h. Yes, 13 − 11 gives the number part of the answer.

i. Problem C: − 26 + 30.
- Are the signs of 26 and 30 the same or different? (Signal.) *Different.*
- So what do you do? (Signal.) *Subtract.* Yes, 30 − 26 gives the number part of the answer.
- Say the problem for the number part of the answer. (Signal.) *30 − 26.*
- (Repeat step i until firm.)

j. Go back to problem A: − 6 minus 9. The signs are the same, so you find the number part of the answer by adding 6 + 9.

- Say the problem for the number part of the answer. (Signal.) *6 + 9.*
- What's the number part of the answer? (Signal.) *15.*
- Write that number after the sign for problem A. √
- (Write on the board:)  [31:2A]

> **a.** − 6 − 9 = − *15*

- Here's what you should have:
  − 6 minus 9 = − 15.
- Everybody, say that fact. (Signal.)
  − *6 − 9 = − 15.*

The same verbal steps are repeated for items b and c, after which students write answers for items d through f independently.

The scripted presentation is designed to help you present the key discriminations quickly and with consistent language, which helps maximize the efficiency of your teaching.

# Language

Some conventions of *Essentials for Algebra* may initially give the impression that the presentations lack integrity because they do not always use the traditional language associated with the content being taught. The reason is simply that what is being taught occurs in stages over many lessons, not all at once in a single lesson or several lessons. The language students need to solve traditional problems will ultimately be taught. The general format of introduction, however, calls for a minimum of vocabulary and a strong emphasis on demonstrating how the operation works, what the discriminations are, and which steps are needed to solve problems. As noted above, vocabulary that is traditionally presented will be introduced later, but vocabulary that is not essential to solving a problem type will probably not be introduced.

# Teaching the Program

*Essentials for Algebra* is designed to be presented to the entire class. The group should be as homogeneous as possible. Students who have similar entry skills and learn at approximately the same rate will progress through the program more efficiently as a group. You should plan to teach a lesson a day. The independent work may be scheduled as homework.

## Organization

Arrange seating so you can receive very quick information on higher performers and lower performers. A good plan is to organize the students something like this:

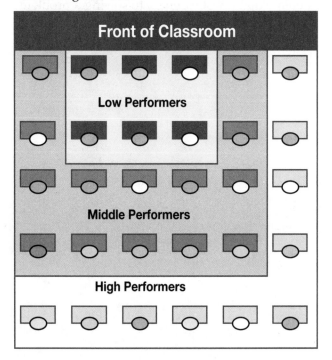

The lowest performers are closest to the front of the classroom. Middle performers are arranged around the lowest performers. Highest performers are arranged around the periphery. With this arrangement, you can position yourself so that you can sample low, average, and high performers by taking a few steps as students work problems.

While different variations of this arrangement are possible, be careful not to seat low performers far from the front center of the room because they require the most feedback. The highest performers, understandably, can be farthest from the center because they attend better, learn faster, and need less observation and feedback.

## Teaching

When you teach the program, a basic rule is that **you should not present from the front of the room unless you are showing something on the board.**

For most of the activities, you direct students to work specified problems. For these activities, you should present from somewhere in the middle of the room (in no set place) and, as students work each problem, observe an adequate sample of students. Although you won't be able to observe every student working every problem, you can observe at least half a dozen students in less than a minute.

**Rehearse any parts of the lesson that are new before presenting the lesson to the class.** Each lesson has a script you will follow. Don't simply read the script, but act it out before you present to the students. Attend to the board displays and how the displays change. If you preview the steps students will take to work the problems in each exercise, you'll be much more fluent in presenting the activity.

**Watch your wording.** Non-board activities are much easier to present than board activities. The board formats are usually designed so they are manageable if you have an idea of the steps you'll take. If you rehearse each of the early lessons before presenting them, you'll soon learn how to present efficiently from the script.

Remind students of the two important rules for doing well in this program: **Always work problems the way they are shown** and **No shortcuts are permitted.**

Remember that everything introduced will be used later. Reinforce students who apply what they learn. Always require students to rework incorrect problems.

# Using the Teacher Presentation Scripts

The script for each lesson indicates precisely how to present each structured activity. The script shows what you say, what you do, and what the students' responses should be.

What you say appears in blue type:
You say this.
What you do appears in parentheses:
(You do this.)
The responses of the students are in italics.
*Students say this.*

Although you may feel uncomfortable "reading" a script (and you may feel that the students will not pay attention), try to present the exercises as if you're saying something important to the students. If you do, you'll find that working from a script is not difficult and that students respond well to what you say.

A sample script appears on page 19. The arrows show five different things you'll do in addition to delivering the wording in the script.

- You'll **signal** to make sure group responses involve all the students. (arrow 1)

- You'll **firm** critical parts of the exercises. (arrow 2)

- You'll **pace** your presentation based on what the students are doing. You'll judge whether to proceed quickly or to wait a few more seconds before moving on with the presentation. (arrow 3)

- You'll **write** (or display) things on the board, and you'll often **change** the board display. (arrow 4)

- You'll **check** students' written work to ensure mastery of the content. (arrow 5)

## ARROW 1: GROUP RESPONSES (SIGNAL.)

Some of the tasks call for group responses. If students respond together with brisk unison responses, you receive good information about whether most of the students are performing correctly. The simplest way to **signal** students to respond together is to adopt a timing practice—just like the timing in a musical piece.

A signal follows a question (as shown by arrow 1) or a direction.

You can signal with a hand drop, clapping one time, snapping your fingers, or tapping your foot. After initially establishing the timing for signals, you can signal through voice inflection and timing.

Students will not be able to initiate responses together at the appropriate rate unless you follow these rules:

a) Talk first. Pause a standard length of time (possibly 1 second); then signal. Never signal while you talk. Don't change the timing from one signal to the next. Students are to respond on your signal—not after it or before it.

b) Model responses that are paced reasonably. Don't permit students to produce slow, droning responses because when students make droning responses, many of them may be copying the responses of others. If students are required to respond at a reasonable speaking rate, all students must initiate responses, and it's relatively easy to determine which students are responding and which students are saying the wrong response. Also, don't permit students to respond at a very fast rate or to "jump" your signal. Listen very carefully to the **first** part of the response.

To correct oral responses that are too fast or too slow, show students exactly what you want them to do. For example:

- My turn to read the equation: 14 plus some number equals 96.
- Your turn: Read the equation. (Signal.) *14 plus some number equals 96.*
- Good reading it the right way.

c) Do not respond with the students unless you are trying to work with them on a difficult response. You present only what's in blue. You do not say the answers with the students, and you should not move your lips or give other spurious clues about the answer.

Think of signals this way: If you use them correctly, they provide you with much diagnostic information. A weak response suggests that you should repeat a task. Signals are, therefore, important early in the program. After students have learned the routine, the students will be able to respond on cue with no signal. That will happen, however, only if you always give your signal at the end of a constant time interval after you complete what you say.

## ARROW 2: FIRMING (REPEAT UNTIL FIRM.)

When students make mistakes, you correct them. A correction may occur during any part of the teacher presentation that calls for the students to respond. It may also occur in connection with what the students are writing.

Here are the rules for correcting mistakes on an oral task:

- **Correct the mistake as soon as you hear it.**

A mistake is either saying the wrong thing or not responding.

- To correct:

  1) **Say the correct answer.**

  2) **Repeat the task the students missed.**

For example:

- Touch point A. √
- How many places to the right of zero is that point? (Signal.) *One place.*

If some students do not respond, respond late, or say anything but *one place*, there's a mistake. As soon as you hear the mistake, correct it.

1) **Say the correct answer.**
   - It's **one** place to the right of zero.
2) **Repeat the task.**
   - How many places to the right of zero is point A? (Signal.) *One.*

Remember, whenever there's a signal, there's a place where students may make mistakes. Correct mistakes as soon as you hear them.

A special correction is needed when correcting mistakes on tasks that teach a relationship. This type of correction is marked with a marginal bracket and the note **(Repeat step __ until firm.)**

The note **(Repeat step __ until firm.)** usually occurs when students must produce more than one related response (as in step b in the sample script).

When you repeat until firm, you follow these steps:

1) **Correct the mistake.** (Tell the answer and repeat the task that was missed.)

2) **Complete the series of tasks included in the bracketed step(s).**

3) **Return to the beginning of the specified step(s) and repeat the entire series of tasks.**

For example, students make a mistake in step b; when the teacher says, "Point to show the direction for the X value," some students don't respond.

1) **Correct the mistake.**

   (Point to the students' right.)
- This is the direction for the X value.
- Point to show the direction for the X value. (Students point right.)

2) **Complete the series of tasks.**

- Point to show the direction for the Y value. (Students point up.)

3) **Repeat step b.**

- Listen again.
- Point to show the direction for the X value. (Students point right.)
- Point to show the direction for the Y value. (Students point up.)

Students show you through their responses whether or not the correction worked, whether or not they are firm.

The repeat-until-firm direction appears only in the most critical parts of new-teaching exercises. It usually focuses on knowledge that is very important for later work. As a general procedure, follow the repeat-until-firm directions. However, if you're quite sure that the mistake was a "glitch" and does not mean that the students lack understanding, ignore the repeat-until-firm direction.

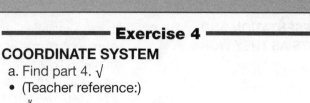

## Exercise 4

### COORDINATE SYSTEM

a. Find part 4. √

• (Teacher reference:)

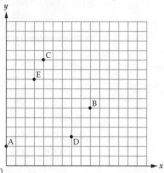

**2** ───────▶
What you firm

b. You're going to write the X and Y values that tell how to reach different points on the coordinate system.
• Point to show the direction for the X value. (Students point right.) √
• Point to show the direction for the Y value. (Students point up.) √
• (Repeat step b until firm.)
c. Touch point A. √ ◀───────── **3**
• How many places to the right of zero is that point? (Signal.) *Zero places.*

How you pace
your presentation

**1** ───────▶
How you secure
group responses

• So X is zero.
• Write that equation and the equation for Y. √
• Check your work.
• (Write on the board:)                          [29: 4A]

| | | |
|---|---|---|
| **A** | $x = 0$, | $y = 2$ |

◀───────── **4**
What you write
(or display)

• Here's what you should have for point A: X = 0, Y = 2.
d. Now write the X and Y equations for the rest of the points. Pencils down when you're finished. (Observe students and give feedback.)
• Check your work.
• (Write to show:)                          [29: 4B]

**5** ───────▶
How you check
students' work

| | | |
|---|---|---|
| **A** | $x = 0$, | $y = 2$ |
| **B** | $x = 9$, | $y = 6$ |
| **C** | $x = 4$, | $y = 11$ |
| **D** | $x = 7$, | $y = 3$ |
| **E** | $x = 3$, | $y = 9$ |

What you add
(or display) next

e. Read both equations for each point.
• Point B. (Signal.) *X = 9, Y = 6.*
  Point C. (Signal.) *X = 4, Y = 11.*
  Point D. (Signal.) *X = 7, Y = 3.*
  Point E. (Signal.) *X = 3, Y = 9.*

## ARROW 3: PACING YOUR PRESENTATION AND INTERACTING WITH STUDENTS AS THEY WORK

You should pace your verbal presentation at a normal speaking rate—as if you were telling somebody something important.

The arrows for number 3 on page 19 show two ways to pace your presentation for activities in which students write or in which they touch or find parts of their workbook or textbook page. The first is a √. The second is a note to **(Observe students and give feedback.)** Both indicate that you will interact with students. Some interactions will serve to correct mistakes. Others reinforce desired behaviors. In other words, √ and **(Observe students and give feedback.)** are signals for managing students and giving feedback that helps the students perform better.

A √ is a note to check what the students are doing. It requires only a second or two. If you are positioned close to several lower-performing students, check whether they are responding appropriately. If they are, proceed with the presentation.

The **(Observe students and give feedback.)** direction implies a more elaborate response. You sample more students and you give feedback, not only to individual students but to the group. Here are the basic rules for what to do and what not to do when you observe and give feedback.

a) Move from the front of the room to a place where you can quickly sample the performance of low, middle, and high performers.

b) As soon as students start to work, start observing. As you observe, make comments to the whole class. Focus these comments on students who are following directions, working quickly, and working accurately. "Wow, a couple of students are almost finished. I haven't seen one mistake so far."

c) Students put their pencils down to indicate that they are finished. Acknowledge students who are finished. They are not to work ahead.

d) If you observe mistakes, do **not** provide a great deal of individual help. Point out any mistake, but do not work the problems for the students. For instance, if a student gets one of the problems wrong, point to it and say, "You

made a mistake." If students don't line up their numerals correctly, say, "You'd better erase that and try again. Your numerals are not lined up." If students are not following instructions that you give, tell them, "You didn't follow my directions. You have to listen carefully. Just work problem A, then stop."

e) If you observe a serious problem that is not unique to the lowest performers, tell the class, "Stop. We seem to have a problem." Point out the mistake. Repeat the part of the exercise that gives them information about what they are to do. **Note:** Do not provide new teaching or new problems. Simply repeat the part of the exercise that gives students the information they need and reassign the work. "Let's see who can get it this time."

f) Do not wait for the slowest students to complete the problems before presenting the workcheck during which students correct their work and fix any mistakes. Allow students a **reasonable** amount of time. You can usually use the middle performers as a gauge for what is reasonable. As you observe that they are completing their work, announce, "Okay, you have about 10 seconds more to finish up." At the end of that time, continue with the exercise.

g) During the workcheck, continue to circulate among the students and make sure they are checking their work. They should fix any mistakes. Praise students who are following the procedure. Allow a reasonable amount of time for them to check each problem. Do not wait for the slowest students to finish their check. Try to keep the workcheck moving as quickly as possible.

h) When higher-performing students do their independent work, you may want to repeat any parts of the lesson with students who had trouble with part of the structured work (made mistakes or didn't finish). Make sure that you check all the problems worked by the lower performers and give them feedback. When you show them what they did wrong, keep your explanation simple. The more involved your explanations, the more likely they are to get confused. If there are serious problems, repeat the exercise that presented difficulties.

If you follow the procedures for observing students and giving feedback, your students will work faster and more accurately. They will also become facile at following your directions.

- If you wait far beyond a reasonable time period before presenting the workcheck, you punish those who worked quickly and accurately. Soon, they will learn that there is no payoff for doing well—no praise, no recognition—but instead a long wait while you give attention to those who are slow.

- If you don't make announcements about students who are doing well and working quickly, the class will not understand what's expected. Students will probably not improve much.

- If you provide extensive individual help on independent work, you will actually reinforce students for not listening to your directions, for being dependent on your help. Furthermore, this dependency becomes contagious. It doesn't take other students in the class long to discover that they don't have to listen to your directions, that they can raise their hand and receive help that shows them how to work the entire problem.

These expectations are the opposite of the ones you want to induce. You want students to be self-reliant and to have **reasons** for learning and remembering what you say when you instruct them. The simplest reasons are that they will use what they have just been shown and that they will receive reinforcement for performing well.

If you follow the management rules outlined above, by the time the students have reached lesson 15, all students should be able to complete assigned work within a reasonable period of time and have reasons to feel good about their ability to do math. That's what you want to happen. Follow the rules, and it will happen. As students improve, you should tell them about it. "What's this? Everybody's finished with that problem already? That's impressive."

## ARROW 4: BOARDWORK/CD

Exercises specify what you are to "write on the board." The teacher materials include a board display CD, which shows all the displays you are to "write on the board." You may use the CD as an alternative for actually writing on the board. To do this, you would project the displays on a screen or wall.

The displays on the CD are labeled consecutively for each lesson. Note that the display code is shown for each "write on the board." The code for the displays shown on page 19 are [29:4A] and [29:4B]. The 29 indicates lesson 29. The 4 indicates that it is exercise 4. The A indicates that it is the first display in that exercise. The B identifies the next display in that exercise. The first display for the next exercise (5) is labeled [29:5A].

If it is not possible for you to use the CD, copy the displays shown in the teacher presentation book display boxes. In the sample exercise, step c, you first write the X and Y values for item A.

- (Write on the board:)                    [29:4A]

| A | $x = 0,$ | $y = 2$ ← |
|---|----------|-----------|

What you write (or display)

In step d, you change the display.

- (Write to show:)                         [29:4B]

| A | $x = 0,$ | $y = 2$ |
|---|----------|---------|
| B | $x = 9,$ | $y = 6$ ← |
| C | $x = 4,$ | $y = 11$ |
| D | $x = 7,$ | $y = 3$ |
| E | $x = 3,$ | $y = 9$ |

What you add (or display) next

Any changes in the original display (additions or alterations) are shown on a white background. The display boxes show both what you'll write and how you'll change the display. If you aren't using the CD, the most efficient way to present the displays is to prepare a transparency of each and show the displays in sequence using an overhead projector.

The best way to use the CD is to stand where the images are projected and direct the presentation from there. Being close to the image allows you to point to details of the display.

Follow these procedures when presenting a new lesson:

1. Select the lesson from the main menu. If you select lesson 25, the screen will show the exercises for lesson 25, starting with the first exercise that has a board display. When you click on an exercise, the display codes will be listed for that exercise.

2. When you reach the first direction to write on the board, click twice on the appropriate code in the menu to display the screen you see in the presentation book.

3. When you come to the next direction for writing on the board, click NEXT, and the next display will appear.

4. Click NEXT to advance to the next display. Click BACK to return to a previous display.

**Note:** The identification code appearing on each CD display corresponds to the code shown in the presentation book.

### ARROW 5: WORKCHECKS (CHECK YOUR WORK.)

It is important to observe students to make sure that they correct their mistakes.

The simplest procedure is for students to use a colored pen for checking. Students should write their work in pencil so they can erase and make any corrections that are necessary as they work. When you indicate that it is time to "Check your work," they put down their pencils and pick up their marking pens. If their work is correct, they mark a **C** for the problem. If their work is wrong, they mark an **X**. Students correct all mistakes before handing in their work. If classwork is part of your grading system, perfect papers and perfectly corrected papers that have no more than 5 errors should receive the same grade.

If you establish organizational procedures at the beginning of the program, students will learn more and learn faster because they will be far more likely to learn from their mistakes. They will not be as tempted to cheat. Instead, they will be more likely to read what they have written on their paper and learn that they will use everything that is introduced in the program.

# Independent Work

The goal of the independent work is to provide review of previously taught work, requiring about 45–60 minutes per lesson. Starting with lesson 11, students are to refer to an answer key when doing their independent work. The key shows the answer to every other item. The key does not show the work. Students check the answer key to see if each answer corresponds. If not, students are to rework the problem to try to obtain the correct answer.

Independent work may be assigned as homework. The number of problems presented for each lesson is relatively high, compared to many traditional programs. The reason for the high number is that if students are brought to mastery on problem types that are taught, they are able to work these problems relatively quickly, without making false starts.

Each newly introduced problem type becomes part of the independent work after it has appeared several times in structured teacher presentations. Everything that is taught in the program becomes part of the independent work. The problems that appear in the independent work are not limited to what is being currently taught. It includes problem types that have been taught earlier.

As a general rule, all major problem types that are taught in the program appear at least 10 times in the independent work. Some appear as many as 30 or more times. Early material is included in later lessons so that the independent work becomes relatively easy for students and provides them with evidence that they are successful.

### UNACCEPTABLE ERROR RATES

Students' independent work should be monitored, and remedies should be provided for error rates that are too high. As a rule, if more than 30 percent of the students miss more than one or two items in any part of the independent work, provide a remedy for that part. In the first lesson in which a recently taught skill is perfectly "independent," error rates for the skill may exceed more than 30 percent. However, if an excessive error rate continues, there is a problem that should be corrected.

High error rates on independent practice may be the result of the following:

a) The students may not be placed appropriately in the program.

b) The initial presentation may not have been adequately firmed. (The students made mistakes that were not corrected. The parts of the teacher presentation in which errors occurred were not repeated until firm.)

c) Students may have received inappropriate help. (When they worked structured problems earlier, they received too much help and became dependent on the help.)

d) Students may not have been required to follow directions carefully.

The simplest remedy for unacceptably high error rates on independent work is to repeat the exercises that occurred immediately before the material became independent. For example, if students have an unacceptable error rate on a particular kind of word problem, go to the last one or two exercises that presented the problem type as a teacher-directed activity. Repeat those exercises until students achieve a high level of mastery. Try to follow the script closely. Make sure you are not providing a great deal of additional prompting. Then assign the independent work for which their error rate was too high. Check to make sure students do not make too many errors.

## Grading Papers and Feedback

The teacher material includes a separate *Answer Key*. The key shows the work for all problems presented during the lesson and as independent work. When students are taught a particular method for working problems, they should follow the steps specified in the key. You should indicate that the work for a problem is wrong if the procedure is not followed.

After completing each lesson and before presenting the next lesson, follow these steps:

1) Make sure **all** errors marked by the students during workchecks have been corrected.

Here's how a corrected mistake should look:

a.
$$
\begin{array}{r} 6\,0^{1}0 \\ -\ 1\,2\,7 \\ \hline 5\,2\,3 \end{array} \quad \mathbf{X}
\qquad
\begin{array}{r} \overset{5\ \ 9}{\cancel{6}}\,\overset{1}{\cancel{0}}\,0 \\ -\ 1\,2\,7 \\ \hline 4\,7\,3 \end{array}
$$

The **X** indicates that the problem was originally wrong. The correct work is shown next to the original problem. This procedure is better than requiring students to erase mistakes. (After they have been erased, you don't know what type of errors students made and, therefore, what type of adjustments would be implied for the next lesson.)

2) Conduct a workcheck for the independent work. One procedure is to provide a structured workcheck of independent work at the beginning of the period. Do not attempt to provide students with complete information about each problem. Read the answers. Students are to mark any mistakes. The workcheck should not take more than five minutes. Students are to correct errors at a later time. You should have some method for checking off each student's name for every lesson. The check shows that the student made corrections and turned in a corrected paper.

3) Spot check each student's corrected paper. Attend to three aspects of the student's work:

a) Were all the mistakes corrected?

b) Is the appropriate work shown for each correction (not just the right answer)?

c) Did the student perform acceptably on tasks that tended to be missed by other students? The answer to this question provides you with information on the student's performance on difficult tasks.

4) Award points for independent work performance. A good plan is to award one point for completing the independent work, one point for correcting all mistakes, and three points for making no more than four errors on the independent work. Students who do well can earn five points for each lesson. These points can be used as part of the basis for assigning grades. The independent work should be approximately one third of the grade. The rest of the grade would be based on the mastery tests. (The independent work would provide students with up to 50 points for a ten-lesson period; each mastery test provides another possible 100 points—100% for a perfect test score.)

# Inducing Appropriate Learning Behaviors

## FOLLOWING DIRECTIONS

Students who place in *Essentials for Algebra* often have strategies for approaching math that are not appropriate. For instance, they may be very poor at following directions, even if they have an understanding of them. For instance, if you instruct the students to work problem **C** and then work problem **A,** a high percentage of them will not follow this direction.

Throughout *Essentials,* you will give students precise directions that may be quite different from those they have encountered earlier in their school experience. For instance, "Write the equation with letters, then stop." Expect a fair percentage of the students to ignore these directions, particularly early in the sequence.

Do not ignore students who do not follow your directions. The most common problems occur when you direct students to work part of a problem and then stop, work one problem and then stop, or write an equation with letters for *all* the problems and then stop.

The simplest remedy for students not following directions is to tell them early in the program that they are to listen to directions and follow them carefully. A good plan is to award points to the group (which means all members of the group earn the same number of points) for following directions. Praise students for attending to them. If you address the issue of following directions early in the school year, students will progress much faster later in the program, and you will not have to nag them about following directions.

What prevents the students who typically place in *Essentials for Algebra* most from learning is their failure to attend to what the teacher says when presenting exercises. The paradox is that the more the teacher tries to explain and to stress important points, the slower the exercise moves, the less students attend, and the less information they remember.

Exercises that are the most difficult for these students (and the most difficult for the teacher to present effectively) are those that have long explanations. An effective procedure is to move fast enough to keep the students attentive.

The general rule is to **go fast** on parts that do not require responses from the students; **go more slowly** and be more emphatic when presenting directions for what the students are to do.

Here's an example from Lesson 55.

---

There are $\frac{3}{5}$ **as many monkeys** as bears.

**The number of monkeys** is $\frac{3}{5}$ the number of bears.

◆ **Rephrase each sentence. Then write a letter equation.**

   **a.** There are $\frac{7}{8}$ as many apples as bananas.

   **b.** There is $\frac{1}{2}$ the number of forks as plates.

---

b. The first statement in the box is part of a new type of algebra word problem: There are 3/5 **as many monkeys** as bears.

- You'll be able to write a correct equation if you say the statement so it starts with the words **the number of.** You can see that statement below the first one:
  **The number of monkeys** is 3/5 the number of bears.

- Remember, start with the words **the number of** and name the first thing named in the sentence—monkeys.

c. I'll read each sentence. You'll say it the other way.

d. Sentence A: There are 7/8 as many apples as bananas.

- What unit is named first in the sentence? (Signal.) *Apples.*

- Start with the words **the number of** and say the whole statement. (Signal.) *The number of apples is 7/8 the number of bananas.*

- (Repeat step d until firm.)

e. Sentence B: There is 1/2 the number of forks as plates.

- Which unit is named first in the sentence? (Signal.) *Forks.*

- Start with the words **the number of** and say the whole statement. (Signal.) *The number of forks is 1/2 the number of plates.*

- (Repeat step e until firm.)

Step b consists of teacher talk with no student responses. Present this part of the script quickly. Don't try to stress the points. Treat the information as something the students can understand and, therefore, something you're summarizing in a perfunctory manner.

Beginning with step c, use a tone of voice that makes the directions very clear.

If you follow the general guidelines of going faster on parts that require no student responses and of providing greater emphasis on parts that check understanding or tell students what they are to do, students will attend better to what you say. They'll learn to listen because they'll learn the link between your explanations and what they'll do.

Here are other guidelines for reinforcing appropriate learning behaviors for students who do not perform well in the first 15 lessons of the program:

1) During the first 15 lessons, hold students to a high criterion of performance. Remind them that they are to follow the procedures you show them.

2) After the first in-program test, which follows lesson 6, provide the specified remedies (see the following section, **In-Program Mastery Tests**); then repeat lessons 4 through 6. Tell students, "We're going to do these lessons again. This time, we'll do them perfectly." Be positive. Reinforce students for following directions and not making the kinds of mistakes they had been making.

After students have completed several lessons perfectly, they will understand your criteria for what they should do to perform acceptably. For some students, the relearning required to perform well is substantial, so be patient, but persistent.

### FOLLOWING SOLUTION STEPS

Another important behavior to address early is that students are to follow the solution steps that are taught. For much of their work, students write equations and rewrite them, possibly two or more times, before solving the equation or answering the question the problem asks. Students who learn the habits of writing and rewriting equations are able to solve problems far more reliably than students who do various calculations without writing equations and then try to relate the calculations.

Do not permit shortcuts or working the problem with steps missing. At first, this convention may strike some students as being laborious. Tell them early in the sequence that if they learn these steps, they will later avoid many difficulties students have when trying to work problems that involve a lot of steps. Point out to students that they will learn "real math strategies" that will permit them to have far less difficulty with higher math than if they were not well versed in solving problems a step at a time.

### LONG LESSONS

Expect some lessons to run long. Do not hurry to complete them in one period if it would realistically take more than the period to present and check the material.

Complete such a lesson during the next period and then start the next lesson during that period and go as far as you can during the allotted time. If you are running long on most of the lessons, the group may not be at mastery, or too much time is being spent on each problem set. The exception is the lesson range 106–120. At least half of these lessons may run long.

## In-Program Mastery Tests

*Essentials for Algebra* provides 11 in-program tests that permit you to assess how well each student is mastering the program content. The first test is in two parts. Test 1A follows lesson 6. Test 1B follows lesson 15. Subsequent tests appear after every tenth lesson, through lesson 115. The primary purpose of the tests is to provide you with information about how well prepared each student is to proceed through the program.

Sometimes students copy from their neighbors. A good method for discovering who is copying is to spread students out during the test if it's physically possible to do so. Discrepancies in the test performance and daily performance of some students pinpoint which students may be copying.

If copying is occurring, reassign seating so that the students who tend to copy are either separated from those who know the answers or are seated near the front center of the room where it is easier for you to monitor them.

Each test has a passing criterion for each part. Remedies are specified for each criterion. Before presenting the next lesson, provide remedies for parts students fail.

Below is test 4, which is scheduled after lesson 45.
Parts 1–4 are in the workbook.

Parts 5–10 are in the textbook.

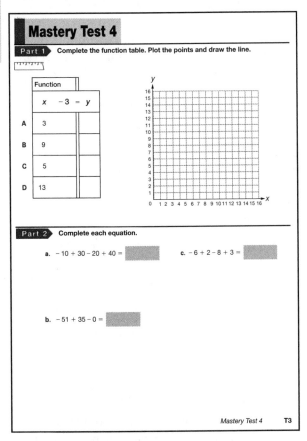

## Mastery Test 4

**Part 1**  Complete the function table. Plot the points and draw the line.

| Function | | |
|---|---|---|
| | $x$  − 3  = $y$ | |
| A | 3 | |
| B | 9 | |
| C | 5 | |
| D | 13 | |

**Part 2**  Complete each equation.

a. $-10 + 30 - 20 + 40 =$ 

c. $-6 + 2 - 8 + 3 =$ 

b. $-51 + 35 - 0 =$ 

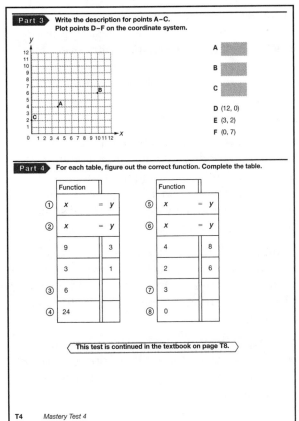

**Part 3**  Write the description for points A–C.
Plot points D–F on the coordinate system.

A 

B 

C 

D (12, 0)

E (3, 2)

F (0, 7)

**Part 4**  For each table, figure out the correct function. Complete the table.

| | Function | | |
|---|---|---|---|
| ① | $x$ | = $y$ | |
| ② | $x$ | = $y$ | |
| | 9 | 3 | |
| | 3 | 1 | |
| ③ | 6 | | |
| ④ | 24 | | |

| | Function | | |
|---|---|---|---|
| ⑤ | $x$ | = $y$ | |
| ⑥ | $x$ | = $y$ | |
| | 4 | 8 | |
| | 2 | 6 | |
| ⑦ | 3 | | |
| ⑧ | 0 | | |

This test is continued in the textbook on page T8.

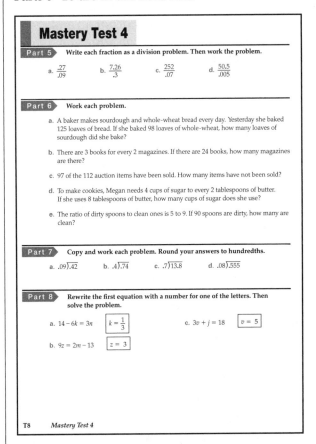

## Mastery Test 4

**Part 5**  Write each fraction as a division problem. Then work the problem.

a. $\frac{.27}{.09}$     b. $\frac{7.26}{.3}$     c. $\frac{252}{.07}$     d. $\frac{50.5}{.005}$

**Part 6**  Work each problem.

a. A baker makes sourdough and whole-wheat bread every day. Yesterday she baked 125 loaves of bread. If she baked 98 loaves of whole-wheat, how many loaves of sourdough did she bake?

b. There are 3 books for every 2 magazines. If there are 24 books, how many magazines are there?

c. 97 of the 112 auction items have been sold. How many items have not been sold?

d. To make cookies, Megan needs 4 cups of sugar to every 2 tablespoons of butter. If she uses 8 tablespoons of butter, how many cups of sugar does she use?

e. The ratio of dirty spoons to clean ones is 5 to 9. If 90 spoons are dirty, how many are clean?

**Part 7**  Copy and work each problem. Round your answers to hundredths.

a. $.09\overline{).42}$     b. $.4\overline{).74}$     c. $.7\overline{)13.8}$     d. $.08\overline{).555}$

**Part 8**  Rewrite the first equation with a number for one of the letters. Then solve the problem.

a. $14 - 6k = 3n$  $\boxed{k = \frac{1}{3}}$     c. $3v + j = 18$  $\boxed{v = 5}$

b. $9z = 2m - 13$  $\boxed{z = 3}$

**Part 9**  Work each problem.

a. The height of the tower is $\frac{3}{2}$ the height of the office building. The office building is 48 feet high. How tall is the tower?

b. The watch costs $83 less than the stereo. The watch costs $61. How much is the stereo?

c. The distance to the school is 2.5 times the distance to the gym. The distance to the gym is 3 miles. What is the distance to the school?

d. Quarterback A threw 361 more yards than quarterback B. Quarterback B threw for 1008 yards. How many yards did quarterback A throw for?

**Part 10**  Write the coordinates for each point on your lined paper.

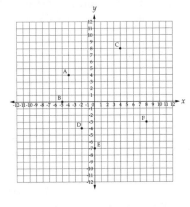

## SCORING AND REMEDIATING THE TESTS

The test section of the answer key provides the correct answer and shows the work for each item.

Tables that accompany each test show the passing score for each part and indicate the percentages for different total test scores.

Here are the tables for test 4.

| TEST 4 PERCENT SUMMARY | | | | | |
|---|---|---|---|---|---|
| Score | % | Score | % | Score | % |
| 65 | 100 | 58 | 89 | 51 | 78 |
| 64 | 98 | 57 | 88 | 50 | 77 |
| 63 | 97 | 56 | 86 | 49 | 75 |
| 62 | 95 | 55 | 85 | 48 | 74 |
| 61 | 94 | 54 | 83 | 47 | 72 |
| 60 | 92 | 53 | 82 | 46 | 71 |
| 59 | 91 | 52 | 80 | | |

| TEST 4 SCORING CHART | | | | | |
|---|---|---|---|---|---|
| Part | Score | | | Possible Score | Passing Score |
| 1 | Function Table | Points | Line | 9 | 7 |
| | 4 | 4 | 1 | | |
| 2 | 2 for each item | | | 6 | 5 |
| 3 | 1 for each description<br>1 for each point | | | 6 | 5 |
| 4 | 1 for each row, 1–8 | | | 8 | 7 |
| 5 | 1 for each item | | | 4 | 3 |
| 6 | Letter Equations b, d, e | Each Answer | | 8 | 7 |
| | 3 | 1 | | | |
| 7 | 1 for each item | | | 4 | 3 |
| 8 | 2 for each item (substitution, answer) | | | 6 | 5 |
| 9 | 2 for each item (letter equation, answer) | | | 8 | 6 |
| 10 | 1 for each point | | | 6 | 5 |
| | | | TOTAL | 65 | |

The scoring chart gives the possible points for each item, the possible points for the part, and the passing criterion.

Students fail a part of the test if they score fewer than the specified number of passing points.

Note that points are sometimes awarded for working different parts of the problem. For example for Part 9, students earn 1 point for the correct letter equation and 1 point for the correct answer.

The percent conversion table shows the percentage grade you'd award students who have a perfect score of 65, a score of 64, and so forth.

## MARKING IN-PROGRAM MASTERY TESTS

Use the criteria in the *Answer Key* for marking each student's test. Record the results on the **Group Summary of Test Performance** (provided on pages 168–173 of this *Teacher's Guide*). The Group Summary of Test Performance can accommodate up to 30 students. The sample below shows only 6 students.

Here's how the results could be summarized following Test 4:

**Remedy Summary—Group Summary of Mastery Test Performance**

***Note:*** Test remedies are specified in the *Answer Key*. Percent Summary is also specified in the *Answer Key*.

| | Test 4 | | | | | | | | | | |
|---|---|---|---|---|---|---|---|---|---|---|---|
| | Check parts not passed | | | | | | | | | | Total % |
| Name | 1 | 2 | 3 | 4 | 5 | 6 | 7 | 8 | 9 | 10 | |
| 1. Amanda Adams | | | √ | | √ | | | | | | 85% |
| 2. William Alberts | | | √ | | | | | | | | 92% |
| 3. Henry Bowman | √ | | √ | | | | | √ | √ | √ | 71% |
| 4. Phillip Caswell | | | | | | | | | | | 89% |
| 5. Zoe Collier | √ | | | | | | | | | | 95% |
| 6. Chan Won Lee | | | √ | | | | | | | √ | 83% |
| Number of students not passed = NP | 2 | 0 | 4 | 0 | 1 | 0 | 0 | 1 | 1 | 2 | |
| Total number of students = T | 6 | 6 | 6 | 6 | 6 | 6 | 6 | 6 | 6 | 6 | |
| Remedy needed if NP/T = 25% or more | Y | N | Y | N | N | N | N | N | N | Y | |

The summary sheet provides you with a cumulative record of each student's performance on the in-program mastery tests.

Summarize each student's performance.

- Make a check in the appropriate columns to indicate any part of the test that was failed.
- At the bottom of each column, write the total number of failures for that part and the total number of students in the class. Then divide the number of failures by the number of students to determine the failure rate for each part.
- Provide a group remedy for each part that has a failure rate of more than 25% (.25).

## TEST REMEDIES

The Answer Key specifies remedies for each test. Any necessary remedies should be presented before the next lesson (lesson 46).

Here are the remedies for Test 4:

> *Note:* Provide any remedies for Test 4 before beginning lesson 46. If more than $\frac{1}{4}$ of the students did not pass a test part, present a remedy for that part. You may reproduce the Remedy Summary Sheet and the blackline masters (Test 4–BLM 1 and BLM 2) in the back of the Teacher's Guide.

### TEST 4 REMEDIES

| Test 4 | Lesson | Exercise | Textbook Part | Test 4 BLM |
|---|---|---|---|---|
| Part 1 | 37 | 2 | 3 and Test 4 BLM–1 | |
| Part 2 | 44 | 5 (Present the items one at a time.) | 2 | – |
| Part 3 | 41 | 5 | – | Test 4 BLM–1 |
| Part 4 | 44 | 2 | – | Test 4 BLM–2 |
| Part 5 | 38 | 4 | 2 | – |
| | 40 | 2 | 2 | – |
| Part 6 | 42 | 5 | 4 | – |
| Part 7 | 43 | 5 | 3 | – |
| Part 8 | 39 | 2 | 1 | – |
| Part 9 | 40 | 1 | 1 | – |
| Part 10 | 44 | 3 | – | Test 4 BLM–2 |

If the same students predictably fail parts of the test, it may be possible to provide remedies for those students as the others do a manageable extension activity. If individual students are weak on a particular skill, they will have trouble later in the program when that skill becomes a component in a larger operation or more complex application.

If students consistently fail tests, they are probably not placed appropriately in the program.

On the completed Group Summary of Test Performance for Mastery Test 4 on page 27, more than 1/4 of the students failed Parts 1, 3, and 10. After you provide group remedies by reteaching the exercises specified as the remedy for those parts, you need to provide individual remedies for Amanda Adams and Henry Bowman because they failed additional parts.

It may not be possible or practical to follow all the steps indicated for correcting individual students who make too many mistakes. You may have the group for only a specified period and have no realistic means of doing much work with individuals or selected students. For these cases, provide test remedies to the entire class and move on in the program, attending to those students who make chronic mistakes to the extent that it is practical without slowing the group's progress.

### BLMs

Note: Before providing the remedies, reproduce the needed BLMs for Test 4 (pages 179 and 180). For this example, page 179 is needed to remedy Parts 1 and 3; page 180 is needed for Parts 4 and 10.

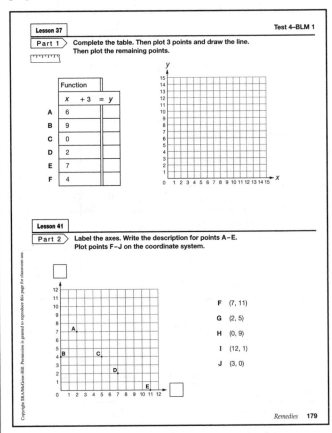

## Sample Lesson

**Note:** Students will need a calculator with a π function for exercise 2.

### ═══ Exercise 1 ═══

**EXPONENTS**

**In Groups**

— **Textbook practice** —

a. Open your textbook to lesson 50, part 1. √
- (Teacher reference:)

$$10 \times 10 \ \times \ 10 \times 10 \times 10 \times 10 \ = 10^6$$
$$(10 \times 10) \times (10 \times 10 \times 10 \times 10) = 10^6$$
$$10^2 \ \times \ 10^4 \ = 10^6$$

$$10 \times 10 \times 10 \ \times \ 10 \times 10 \times 10 \ = 10^6$$
$$(10 \times 10 \times 10) \times (10 \times 10 \times 10) = 10^6$$
$$10^3 \ \times \ 10^3 \ = 10^6$$

- You've learned how to express repeated multiplication as a base and exponent.
b. The first equation shows a set of 10s.
- What's the base? (Signal.) *10.*
  The base is 10. The base is shown 6 times.
- So what's the exponent? (Signal.) *6.*
  So the whole set is $10^6$.
c. Below is the same set of 10s in 2 groups. The groups are **multiplied** together. How many 10s are multiplied in the first group? (Signal.) *2.*
  So that group equals $10^2$.
  Say the base and exponent for that group. (Signal.) *$10^2$.*
- Look at the next group. √
  How many 10s are in the second group? (Signal.) *4.*
- Say the base and exponent for that group. (Signal.) *$10^4$.*
  So another way to show $10^6$ is $10^2$ times $10^4$.
- What's another way of showing $10^6$? (Signal.) *$10^2 \times 10^4$.*
- (Repeat step c until firm.)
d. The next box shows the same set of 10s in different groups.
- How many 10s are in the first group in parentheses? (Signal.) *3.*

Say the base and exponent for that group. (Signal.) *$10^3$.*
- How many 10s are in the other group? (Signal.) *3.*
  Say the base and the exponent for that group. (Signal.) *$10^3$.*
- So $10^3 \times 10^3 = 10^6$.
- What's another way of showing $10^6$? (Signal.) *$10^3 \times 10^3$.*
e. So if the base number is shown 6 times, the exponents must add up to 6.
f. If the base is shown 6 times, what must the exponents add up to? (Signal.) *6.*
- If the base is shown 9 times, what must the exponents add up to? (Signal.) *9.*
- If the base is shown 5 times, what must the exponents add up to? (Signal.) *5.*
- (Repeat step f until firm.)

— **Textbook practice** —

a. Find part 2. √
- For each item, you'll write the complete equation with exponents.
b. Problem A. The multiplication shows 8 seven times.
- Say the base and exponent for all the 8s. (Signal.) *$8^7$.*
  So no matter how the 8s are multiplied together, the exponents must add up to 7.
- You can see the groups set off with parentheses.
- Touch the first group. √
  Tell me the base and exponent you'll write for the first group. (Signal.) *$8^2$.*
- Next group.
  Tell me the base and exponent. (Signal.) *$8^3$.*
- Last group.
- Tell me the base and exponent. (Signal.) *$8^2$.*
- The exponents are 2 and 3 and 2. Do the exponents add up to 7? (Signal.) *Yes.*
- So the whole equation is $8^7 = 8^2 \times 8^3 \times 8^2$.
c. Say the equation. (Signal.) $8^7 = 8^2 \times 8^3 \times 8^2$.
- Write that equation. Pencils down when you're finished. √

- (Write on the board:)        [50:1A]

   **a.**     $8^7 = 8^2 \times 8^3 \times 8^2$

- Here's what you should have.
- d. Write the complete equation for problem B. Pencils down when you're finished. (Observe students and give feedback.)
- (Write on the board:)        [50:1B]

   **b.**     $7^5 = 7^3 \times 7^2$

- Here's what you should have.
- e. Write the complete equation for the rest of the items in part 2. Pencils down when you're finished. (Observe students and give feedback.)
- f. Check your work. Read each equation.
- Equation C. (Signal.) $9^9 = 9^4 \times 9^2 \times 9^3$.
- Equation D. (Signal.) $5^4 = 5^2 \times 5^2$.
- Equation E. (Signal.) $10^8 = 10^3 \times 10^3 \times 10^2$.
- g. Raise your hand if you got everything right. √

---
**Exercise 2**
---

**CIRCUMFERENCE/DIAMETER**

**— Textbook practice —**

- a. Find part 3. √
- b. You're going to work problems that start with the equation for the circumference of a circle.
- What's the name for 3.14? (Signal.) *Pi.*
- Say the equation for the circumference of a circle. (Signal.) $C = \pi D$.
- For some problems, you'll find the diameter. For others, you'll find the circumference.
- c. Touch circle A. √
- What is given, the circumference or the diameter? (Signal.) *Circumference.*
- So you solve for the diameter.
- What do you solve for? (Signal.) *Diameter.*
- d. Circle B. What is given, the circumference or the diameter? (Signal.) *Diameter.*

- So what do you solve for? (Signal.) *Circumference.*
- e. Circle C. What is given? (Signal.) *Circumference.*
- So what do you solve for? (Signal.) *Diameter.*
- f. Circle D. What is given? (Signal.) *Diameter.*
- So what do you solve for? (Signal.) *Circumference.*
- g. Circle E. What is given? (Signal.) *Circumference.*
- So what do you solve for? (Signal.) *Diameter.*
- h. Work problem A. Use the $\pi$ key on your calculator. Pencils down when you're finished. (Observe students and give feedback.)
- (Write on the board:)        [50:2A]

   **a.**        $C = \pi d$

$$\left(\frac{1}{\pi}\right) 11 = \pi d \left(\frac{1}{\pi}\right)$$

$$\frac{11}{\pi} = d$$

$$\boxed{3.50 \ m}$$

- Here's what you should have.
- The circumference is 11 meters. What problem did you work on your calculator? (Signal.) $11 \div \pi$.
- What's the diameter? (Signal.) *3.50 meters.*
- i. Work problem B. Pencils down when you're finished. (Observe students and give feedback.)
- (Write on the board:)        [50:2B]

   **b.**        $C = \pi d$
                   $C = \pi (4.5)$

$$\boxed{14.14 \ yd}$$

- Here's what you should have.
- The diameter is 4.5 yards.
- What problem did you work on your calculator? (Signal.) $\pi \times 4.5$.
- What's the circumference? (Signal.) *14.14 yards.* [*14.13* if 3.14 is used.]

j. Work the rest of the problems in part 3. Pencils down when you're finished. (Observe students and give feedback.)

k. Check your work.

l. Problem C. The circumference is 2.08 feet.

- What problem did you work on your calculator? (Signal.) *2.08 ÷ π.*
- What's the diameter? (Signal.) *0.66 feet.*

m. Problem D. The diameter is 29 inches.

- What problem did you work on your calculator? (Signal.) *π × 29.*
- What's the circumference? (Signal.) *91.11 inches. [91.06 if 3.14 is used.]*

n. Problem E. The circumference is 0.8 centimeters.

- What problem did you work on your calculator? (Signal.) *.8 ÷ π.*
- What's the diameter? (Signal.) *0.25 centimeters.*

## Exercise 3

### RATE EQUATIONS
### Reverse Order

#### — Textbook practice —

a. Find part 4. √

- These are problems you solve with rate equations.
- Last time you wrote the equations so they start with the unit that answers the question.

b. Problem A: A machine produces pencils at the rate of 120 pencils per minute. How long will it take to produce 40 pencils?

- Raise your hand when you know which unit the problem asks about. √
- Which unit? (Signal.) *Minutes.*
- (Write on the board:) [50:3A]

$$\textbf{a.} \quad m = m$$

- Start with the simple equation **M = M,** and complete the rate equation. Pencils down when you've done that much. (Observe students and give feedback.)
- Check your work.

- (Write to show:) [50:3B]

$$\textbf{a.} \quad m = \left(\frac{m}{p}\right) p$$

- Here's what you should have: M = M over P times P.

c. Problem B: There are 3.5 pounds of flour for every pound of sugar. How many pounds of flour are used if 10 pounds of sugar are used?

- Tell me which unit the problem asks about. (Pause. Signal.) *Pounds of flour.*
- Skip 5 lines. Start with the simple equation **PF = PF,** and complete the rate equation. Pencils down when you're finished. (Observe students and give feedback.)
- Check your work.
- (Write on the board:) [50:3C]

$$\textbf{b.} \quad pf = \left(\frac{pf}{ps}\right) ps$$

- Here's what you should have: PF = PF over PS times PS.

d. Write letter equations for problems in C and D. Leave space below each equation. Pencils down when you've done that much. (Observe students and give feedback.)

- Problem C. Read the equation that begins with W. (Signal.) *W = (W/M) M.*
- Problem D. Read the equation that begins with CM. (Signal.) *CM = (CM/Y) Y.*

e. Now work all the problems in part 4. Answer each question with a number and a unit name. Pencils down when you're finished. (Observe students and give feedback.)

f. Check your work.

- Problem A. How long will it take to produce 40 pencils? (Signal.) *1/3 minute.*
- Problem B. How many pounds of flour are used? (Signal.) *35 pounds.*
- Problem C. How many women work in the factory? (Signal.) *160 women.*

- Problem D. How much will the diameter increase? (Signal.) *18 and 2/3 centimeters.*

---
## Exercise 4
---

## MULTIPLYING INTEGERS

**— Textbook practice —**

a. Find part 5. √
- These are multiplication problems with signed numbers.
b. Remember the rules for multiplying 2 values.
- If the signs are the same, what is the sign in the answer? (Signal.) *Plus.*
- If the signs are different, what is the sign in the answer? (Signal.) *Minus.*
- (Repeat step b until firm.)
c. Everybody, read problem A. (Signal.) *– 5 (– 2.3).*
- Are the signs the same or different? (Signal.) *Same.*
- So what's the sign in the answer? (Signal.) *Plus.*
d. Read problem B. (Signal.) *– 3/8 (+ 5).*
- Are the signs the same or different? (Signal.) *Different.*
- So what's the sign in the answer? (Signal.) *Minus.*
e. Copy the problems in part 5 and work them.
- Remember, first figure out the sign in the answer. Then multiply to find the number part of the answer. Pencils down when you're finished.
  (Observe students and give feedback.)
f. Check your work.
- Problem A: – 5 (– 2.3).
  What's the answer? (Signal.) *+ 11.5.*
- Problem B: – 3/8 (+ 5).
  What's the answer? (Signal.) *– 15/8.*
- Problem C: + 6.4 (– 10).
  What's the answer? (Signal.) *– 64.*
- Problem D: – .4 (+ 2).
  What's the answer? (Signal.) *– .8.*
- Problem E: – 7 (– 1).
  What's the answer? (Signal.) *+ 7.*
- Problem F: – 5/7 (– 6).
  What's the answer? (Signal.) *+ 30/7.*

- Problem G: + 1 (– 6).
  What's the answer? (Signal.) *– 6.*
- Problem H: – 2/3 (+ 7).
  What's the answer? (Signal.) *– 14/3.*

---
## Exercise 5
---

## ALGEBRA
## Like Terms on Both Sides

**— Textbook practice —**

a. Find part 6. √
b. Problem A: $9W - 3W = 10 + W - 4$.
- Remember the steps: First, combine like terms on each side. Then add or subtract to get a letter term on 1 side and a number term on the other side. Then solve for the letter. Pencils down when you've finished problem A.
  (Observe students and give feedback.)
- (Write on the board:) [50:5A]

$$
\begin{aligned}
9w - 3w &= 10 + w - 4 \\
6w &= 6 + w \\
-w &\quad\; -w \\
\hline
\left(\tfrac{1}{5}\right) 5w &= 6 \left(\tfrac{1}{5}\right) \\
\end{aligned}
$$

$$\boxed{w = \frac{6}{5}}$$

- The equation with combined like terms is $6W = 6 + W$.
- You subtract W from both sides. You get the equation $5W = 6$. So $W = 6/5$.
c. Problem B: $4R - 1 - 13 - R = 3 + 4$.
- Combine the like terms. Then solve for R. Pencils down when you're finished.
  (Observe students and give feedback.)
- (Write on the board:) [50:5B]

$$
\begin{aligned}
4r - 1 - 13 - r &= 3 + 4 \\
3r - 14 &= 7 \\
+ 14 &\quad + 14 \\
\hline
\left(\tfrac{1}{3}\right) 3r &= 21 \left(\tfrac{1}{3}\right) \\
\end{aligned}
$$

$$\boxed{r = 7}$$

- Read the equation with combined like terms. (Signal.) *3R − 14 = 7.*
- What do you do to change both sides? (Signal.) *Add 14.*
  So 3R = 21.
- What does R equal? (Signal.) *7.*

d. Problem C: 10 − 2 = 2 thirds H + 6 + 5 thirds H.

- Combine the like terms. Then figure out what H equals. Pencils down when you're finished.
  (Observe students and give feedback.)
- (Write on the board:) [50:5C]

$$
\begin{array}{c}
\text{c.} \quad 10 - 2 = \dfrac{2}{3}h + 6 + \dfrac{5}{3}h \\[6pt]
8 = \dfrac{7}{3}h + 6 \\[4pt]
\underline{-6 \qquad\qquad -6} \\[4pt]
\left(\dfrac{3}{7}\right) 2 = \dfrac{7}{3}h \left(\dfrac{3}{7}\right) \\[6pt]
\boxed{\dfrac{6}{7} = h}
\end{array}
$$

- Read the equation with combined like terms. (Signal.) *8 = 7 thirds H + 6.*
- What do you do to change both sides? (Signal.) *Subtract 6.*
- What does H equal? (Signal.) *6/7.*

e. Problem D: 11K − 4K = 15 + 2K − 5.

- Combine the like terms. Then figure out what K equals. Pencils down when you're finished.
  (Observe students and give feedback.)
- (Write on the board:) [50:5D]

$$
\begin{array}{c}
\text{d.} \quad 11k - 4k = 15 + 2k - 5 \\[4pt]
7k = 10 + 2k \\[4pt]
\underline{-2k \qquad\quad -2k} \\[4pt]
\left(\dfrac{1}{5}\right) 5k = 10 \left(\dfrac{1}{5}\right) \\[6pt]
\boxed{k = 2}
\end{array}
$$

- Read the equation with combined like terms. (Signal.) *7K = 10 + 2K.*
- What do you do to change both sides? (Signal.) *Subtract 2K.*

- What does K equal? (Signal.) *2.*

f. Problem E: 3G − 7G − 10 + 40 = G.

- Combine the like terms. Then figure out what G equals. Pencils down when you're finished.
  (Observe students and give feedback.)
- (Write on the board:) [50:5E]

$$
\begin{array}{c}
\text{e.} \quad 3g - 7g - 10 + 40 = g \\[4pt]
-4g \qquad\qquad + 30 = g \\[4pt]
\underline{+ 4g \qquad\qquad\qquad + 4g} \\[4pt]
\left(\dfrac{1}{5}\right) 30 = 5g \left(\dfrac{1}{5}\right) \\[6pt]
\boxed{6 = g}
\end{array}
$$

- Read the equation with combined like terms. (Signal.) *− 4G + 30 = G.*
- What do you do to change both sides? (Signal.) *Add 4G.*
- What does G equal? (Signal.) *6.*

━━━━━ **Exercise 6** ━━━━━
**INDEPENDENT WORK**

**Assign Independent Work: textbook parts 7−12 and workbook parts 1 and 2.**

## Part 1 — Exponents for Multiplied Groups

$$10 \times 10 \times 10 \times 10 \times 10 \times 10 = 10^6$$
$$(10 \times 10) \times (10 \times 10) \times (10 \times 10) = 10^6$$
$$10^2 \times 10^4 = 10^6$$

$$10 \times 10 \times 10 \times 10 \times 10 \times 10 = 10^6$$
$$(10 \times 10 \times 10) \times (10 \times 10 \times 10) = 10^6$$
$$10^3 \times 10^3 = 10^6$$

♦ If the base is shown 6 times, the exponents must add up to 6.

## Part 2

**Copy and complete each boxed equation.**

a. $(8 \times 8) \times (8 \times 8 \times 8) \times (8 \times 8)$

$8^7 = \blacksquare \times \blacksquare \times \blacksquare$

b. $(7 \times 7 \times 7) \times (7 \times 7)$

$7^5 = \blacksquare \times \blacksquare$

c. $(9 \times 9 \times 9 \times 9) \times (9 \times 9) \times (9 \times 9 \times 9)$

$9^9 = \blacksquare \times \blacksquare \times \blacksquare$

d. $(5 \times 5) \times (5 \times 5)$

$5^4 = \blacksquare \times \blacksquare$

e. $(10 \times 10 \times 10) \times (10 \times 10 \times 10) \times (10 \times 10)$

$10^8 = \blacksquare \times \blacksquare \times \blacksquare$

## Part 3

**Find the circumference or the diameter of each circle.**

a.
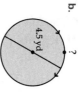
11 m

b.

4.5 yd

c.
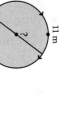
2.08 ft

d.

29 in.

e.

0.8 cm

## Part 4

**For each problem, write the rate equation with letters. Start with the letter that answers the question. Then work the problem.**

a. A machine produces pencils at the rate of 120 pencils per minute. How long will it take to produce 40 pencils?

b. There are 3.5 pounds of flour for every pound of sugar. How many pounds of flour are used if 10 pounds of sugar are used?

c. How many women work in a factory that employs 96 men if the ratio of men to women is 3 to 5?

d. A tree trunk grows at a steady rate. If the diameter of the tree trunk increases by 8 centimeters in 3 years, by how much will the diameter increase in 7 years?

## Part 5

**Copy and work each item. First figure out the sign in the answer. Then multiply.**

a. $-5(-2.3) = \blacksquare$

b. $-\frac{3}{8}(+5) = \blacksquare$

c. $+6.4(-10) = \blacksquare$

d. $-.4(+2) = \blacksquare$

e. $-7(-1) = \blacksquare$

f. $-\frac{5}{7}(-6) = \blacksquare$

g. $+1(-6) = \blacksquare$

h. $-\frac{2}{3}(+7) = \blacksquare$

**Part 6**  Copy each equation. Rewrite each equation with like terms combined. Then solve for the letter.

**a.** $9w - 3w = 10 + w - 4$

**b.** $4r - 1 - 13 - r = 3 + 4$

**c.** $10 - 2 = \frac{2}{3}h + 6 + \frac{5}{3}h$

**d.** $11k - 4k = 15 + 2k - 5$

**e.** $3g - 7g - 10 + 40 = g$

**Part 7**  Work each problem.

**a.** The jeans cost $11.30 less than the coat. The jeans cost $17.55. How much did the coat cost?

**b.** There were 63 people in the park. 17 were on vacation. How many were not on vacation?

**c.** Jon's shoes were 1.3 inches shorter than Eric's shoes. Jon's shoes were 10.5 inches long. How long were Eric's shoes?

**d.** There are 72 boys and 59 girls at the show. How many children are at the show?

**e.** The number of tables is $\frac{2}{3}$ the number of people. There are 42 tables. How many people are there?

**Part 8**  Complete each equation to show the base and exponent.

**a.** $8 \times 8 \times 8 \times 8 \times 8 =$ ■

**b.** $10 \times 10 \times 10 =$ ■

**Part 9**  Solve each problem.

**a.** $2t - 30m = 20$

$\boxed{m = \frac{1}{5}}$

**b.** $14 + m = 2k$

$\boxed{m = 12}$

**c.** $\frac{1}{8}r + t = 1$

$\boxed{r = -3}$

**Part 10**  Work each problem.

**a.** $30 - 5 = -\frac{3}{8}p + p$

**b.** $2b - 2 + 10b - 1 = 3$

**c.** Red and yellow birds are nesting in the tree. There are 26 yellow birds. If 48 birds are nesting in the tree, how many red birds are there?

**d.** A shadow is $\frac{5}{7}$ the height of the hill. The hill is 56 feet high. How long is the shadow?

**e.** The dog is 2.6 months younger than the cat. The cat is 20.8 months old. What's the age of the dog?

**f.** Furnace M is 66 degrees hotter than furnace P. If furnace P is 588 degrees, what is the temperature inside furnace M?

**g.** 12 of the children are sick. The rest are well. There are 135 children. How many are well?

**Part 11**  Find the area and perimeter of each figure.

**a.**

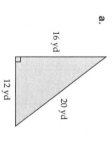

16 yd    20 yd    12 yd

**b.**

14.3 ft

5.2 ft

**Part 12**  Copy and complete each equation. First show the multiplication. Then show the value it equals.

**a.** $5^4 =$ ■  $=$ ■

**b.** $7^5 =$ ■  $=$ ■

**c.** $.5^4 =$ ■  $=$ ■

**d.** $14^3 =$ ■  $=$ ■

## Lesson 50

### Independent Work

**Part 1** ▶ Complete the table.

| Function | x | $\left(\dfrac{y}{x}\right) = y$ |
|---|---|---|
| A | 4 | 5 |
| B | | 25 |
| C | 12 | |
| D | | 10 |
| E | 28 | |

**Part 2** ▶ Round each value to a whole number, 10ths, and 100ths.

| | Whole Number | Tenths | Hundredths |
|---|---|---|---|
| 3.546 | | | |
| 2.099 | | | |
| 6.340 | | | |

---

## Lesson 50

### Independent Work

**Part 1** ▶ Complete the table.

| Function | x | $\left(\dfrac{y}{x}\right) = y$ |
|---|---|---|
| A | 4 | 5 |
| B | 20 | 25 |
| C | 12 | 15 |
| D | 8 | 10 |
| E | 28 | 35 |

$B \quad \left(\dfrac{4}{5}\right) x \left(\dfrac{5}{4}\right) = 25 \left(\dfrac{4}{5}\right) \quad \boxed{x = 20}$

$D \quad \left(\dfrac{4}{5}\right) x \left(\dfrac{5}{4}\right) = 10 \left(\dfrac{4}{5}\right) \quad \boxed{x = 8}$

**Part 2** ▶ Round each value to a whole number, 10ths, and 100ths.

| | Whole Number | Tenths | Hundredths |
|---|---|---|---|
| 3.546 | 4 | 3.5 | 3.55 |
| 2.099 | 2 | 2.1 | 2.10 |
| 6.340 | 6 | 6.3 | 6.34 |

## Lesson 50 Textbook

**Part 2**

a. $8^7 = \boxed{8^2} \times \boxed{8^5}$

b. $7^5 = \boxed{7^3} \times \boxed{7^2}$

c. $9^9 = \boxed{9^4} \times \boxed{9^2} \times \boxed{9^3}$

d. $5^4 = \boxed{5^2} \times \boxed{5^2}$

e. $10^8 = \boxed{10^3} \times \boxed{10^3} \times \boxed{10^2}$

**Part 3**

a. $C = \pi d$
$(\frac{1}{\pi})\,11 = \pi d\,(\frac{1}{\pi})$
$\frac{11}{\pi} = d$
$\boxed{3.50\,m}$

b. $C = \pi d$
$C = \pi (4.5)$
$\boxed{14.14\,yd}$  **Without π: 14.13 yd**

c. $C = \pi d$
$(\frac{1}{\pi})\,2.08 = \pi d\,(\frac{1}{\pi})$
$\frac{2.08}{\pi} = d$
$\boxed{.66\,ft}$

d. $C = \pi d$
$C = \pi (29)$
$\boxed{91.11\,in.}$  **Without π: 91.06 in.**

**Part 4**

a. $m = (\frac{m}{p})\,P$
$m = (\frac{1}{120})\,40$
$m = \frac{1}{3}$
$\boxed{\frac{1}{3}\,minute}$

b. $pf = (\frac{pf}{ps})\,ps$
$pf = (\frac{5.5}{1})\,10$
$pf = 55$
$\boxed{55\,pounds\,[of\,flour]}$

c. $w = (\frac{w}{m})\,m$
$w = (\frac{5}{3})\,96$
$w = 160$
$\boxed{160\,women}$

d. $cm = (\frac{cm}{y})\,y$
$cm = (\frac{8}{3})\,7$
$cm = \frac{56}{3}$
$\boxed{18\frac{2}{3}\,cm}$

**Part 5**

a. $-5(-2.3) = \boxed{+11.5}$

b. $-\frac{3}{8}(+5) = \boxed{-\frac{15}{8}}$

c. $+6.4(-10) = \boxed{-64.0}$

d. $-.4(+2) = \boxed{-.8}$

e. $-7(-1) = \boxed{+7}$

f. $-\frac{5}{7}(-6) = \boxed{+\frac{30}{7}}$

g. $+1(-6) = \boxed{-6}$

h. $-\frac{2}{3}(+7) = \boxed{-\frac{14}{3}}$

105

## Lesson 50 Part 6

a. $9w - 3w = 10 + w - 4$
$6w = 6 + w$
$\quad\ -w \quad\quad -w$
$(\frac{1}{5})\,5w = 6\,(\frac{1}{5})$
$\boxed{w = \frac{6}{5}}$

b. $4r - 1 - 13 - r = 3 + 4$
$3r - 14 = 7$
$\quad\ +14 \quad +14$
$(\frac{1}{3})\,3r = 21\,(\frac{1}{3})$
$\boxed{r = 7}$

c. $10 - 2 = \frac{2}{3}h + 6 + \frac{5}{3}h$
$8 = \frac{7}{3}h + 6$
$\quad -6 \quad\quad -6$
$(\frac{3}{7})\,2 = \frac{7}{3}h\,(\frac{3}{7})$
$\boxed{\frac{6}{7} = h}$

d. $11k - 4k = 15 + 2k - 5$
$7k = 10 + 2k$
$\quad -2k \quad\quad -2k$
$(\frac{1}{5})\,5k = 10\,(\frac{1}{5})$
$\boxed{k = 2}$

e. $5g - 7g - 10 + 40 = g$
$-4g + 30 = g$
$+4g \quad\quad +4g$
$(\frac{1}{5})\,30 = 5g\,(\frac{1}{5})$
$\boxed{6 = g}$

## Lesson 50 Independent Work

**Part 7**

a.
$j = c - 11.30$
$17.55 = c - 11.30$
$+11.30 \quad +11.30$
$28.85 = c$
$\boxed{\$28.85}$

b.
$P = o + no$
$63 = 17 + no$
$-17 \quad -17$
$46 = no$
$\boxed{46\ people\ [not\ on\ vacation]}$

c.
$j = E - 1.3$
$10.5 = E - 1.3$
$+1.3 \quad +1.3$
$11.8 = E$
$\boxed{11.8\ inches}$

d.
$c = b + g$
$c = 72 + 59$
$c = 151$
$\boxed{151\ children}$

e.
$t = \frac{2}{3}P$
$(\frac{3}{2})\,42 = \frac{2}{3}P\,(\frac{3}{2})$
$63 = P$
$\boxed{63\ people}$

**Part 8**

a. $8 \times 8 \times 8 \times 8 \times 8 = \boxed{8^5}$

b. $10 \times 10 \times 10 = \boxed{10^3}$

106

a. $2t - 30m = 20$
$2t - 30(\frac{1}{5}) = 20$
$2t - 6 = 20$
$\phantom{2t} + 6 \quad + 6$
$(\frac{1}{2})2t = 26(\frac{1}{2})$
$\boxed{t = 13}$

b. $14 + m = 2k$
$14 + 12 = 2k$
$(\frac{1}{2})26 = 2k(\frac{1}{2})$
$\boxed{13 = k}$

$\boxed{m = 12}$

$\boxed{m = \frac{1}{5}}$

c. $\frac{1}{8}r + t = 1$
$\frac{1}{8}(-3) + t = 1$
$-\frac{3}{8} + t = 1$
$\phantom{-} + \frac{3}{8} \quad + \frac{3}{8}$
$t = \frac{11}{8}$

$\boxed{r = -3}$

$\boxed{\text{or } t = 1\frac{3}{8}}$

**Part 10**

a. $30 - 5 = -\frac{3}{8}p + p$
$25 = -\frac{3}{8}p + \frac{8}{8}p$
$25 = \frac{5}{8}P$
$(\frac{8}{5})25 = \frac{5}{8}P(\frac{8}{5})$
$\boxed{40 = P}$

b. $2b - 2 + 10b - 1 = 3$
$12b - 3 = 3$
$\phantom{12b} + 3 \quad + 3$
$(\frac{1}{12})12b = 6(\frac{1}{12})$
$\boxed{b = \frac{1}{2}}$

c. $b = r + y$
$48 = r + 26$
$- 26 \quad - 26$
$22 = r$
$\boxed{22 \text{ red birds}}$

d. $s = 5h$
$s = \frac{5}{7}(56)$
$s = 40$
$\boxed{40 \text{ feet}}$

e. $d = c - 2.6$
$d = 20.8 - 2.6$
$d = 18.2$
$\boxed{18.2 \text{ months old}}$

f. $M = P + 66$
$M = 588 + 66$
$M = 654$
$\boxed{654 \text{ degrees}}$

g. $c = s + w$
$135 = 12 + w$
$- 12 \quad - 12$
$123 = w$
$\boxed{123 \text{ [well] children}}$

**Part 11**

a. $A = \frac{1}{2}(b \times h)$
$A = \frac{1}{2}(12 \times 16)$
$A = \frac{1}{2}(192)$
$\boxed{96 \text{ sq yd}}$

$P = 12$
$\phantom{P =} 16$
$+ 20$
$\boxed{48 \text{ yd}}$

b. $A = b \times h$
$A = 14.3 \times 5.2$
$\boxed{74.36 \text{ sq ft}}$

$P = 14.3$
$\phantom{P =} 5.2$
$\phantom{P =} 14.3$
$+ \phantom{1}5.2$
$\boxed{39.0 \text{ ft}}$

**Part 12**

a. $5^4 = 5 \times 5 \times 5 \times 5 = \boxed{625}$

b. $7^5 = 7 \times 7 \times 7 \times 7 \times 7 = \boxed{16{,}807}$

c. $.5^4 = .5 \times .5 \times .5 \times .5 = \boxed{.0625}$

d. $14^3 = 14 \times 14 \times 14 = \boxed{2744}$

## Sample Lesson

━━━━━━━━━━ **Exercise 1** ━━━━━━━━━━
### SCIENTIFIC NOTATION
━ **Workbook practice** ━

a. Open your workbook to lesson 92, part 1. √
- (Teacher reference:)

| | |
|---|---|
| | $50{,}300 = \blacksquare \times 10^{\blacksquare}$ |
| ◆ Copy the digits before the final zeros. | $5\ 03 \times 10^{\blacksquare}$ |
| ◆ Write a decimal point after the first digit. | $5.03 \times 10^{\blacksquare}$ |
| ◆ Write the exponent for 10. | $5.03 \times 10^{4}$ |

b. The box shows how to rewrite a number as something multiplied by 10 to some power.
- Here's how you do that. The number is 50,300. First you copy the digits that are before the final zeros. That's 5, zero, 3.
- Which digits do you copy? (Signal.) *5, 0, 3.*
- You write a decimal point after the first digit.
- What's the first digit? (Signal.) *5.*
  So you write a decimal point after the 5.
- What decimal value do you write in the first box? (Signal.) *5.03.*
- Then you figure out the exponent of 10.
- What's the exponent of 10? (Signal.) *4.*
  You write the exponent in the other box.
- 50,300 equals 5.03 times $10^4$.

━ **Workbook practice** ━

a. Find part 2 of your workbook. √
b. Problem A: 752,000.
- What are the digits before the zeros? (Signal.) *7, 5, 2.*
- So what decimal value do you write in the first box? (Signal.) *7.52.*
c. Problem B: 7,131,000.
- What are the digits before the zeros? (Signal.) *7, 1, 3, 1.*
- So what decimal value do you write in the first box? (Signal.) *7.131.*
d. Problem C: 37,000.
- What are the digits before the zeros? (Signal.) *3, 7.*
- So what decimal value do you write in the first box? (Signal.) *3.7.*

e. Problem D: 8040.
- What are the digits before the final zero? (Signal.) *8, 0, 4.*
- So what decimal value do you write in the first box? (Signal.) *8.04.*
f. Problem E: 70,180,000.
- What are the digits before the final zeros? (Signal.) *7, 0, 1, 8.*
- So what decimal value do you write in the first box? (Signal.) *7.018.*
g. After you write the decimal value in the first box, you write the exponent for 10.
- You just start after the first digit of the original value and write the number of places to the end of the original value.
h. (Write on the board:)                    [92:1A]

| | | |
|---|---|---|
| **a.** | $752{,}000 =$ ☐ | $\times 10^{\square}$ |

- Here's problem A. Tell me the decimal value that goes in the first box. (Signal.) *7.52.*
- (Write to show:)                    [92:1B]

| | | |
|---|---|---|
| **a.** | $752{,}000 = \boxed{7.52}$ | $\times 10^{\square}$ |

- Now I start after the first digit of 752,000 and count the places to the end of the number.
- Raise your hand when you know the number of places. √
- Everybody, how many places? (Signal.) *5.*
- So the exponent for 10 is 5.
- (Write to show:)                    [92:1C]

| | | |
|---|---|---|
| **a.** | $752{,}000 = \boxed{7.52}$ | $\times 10^{\boxed{5}}$ |

- 752,000 equals 7.52 times $10^5$.
i. Your turn: Complete item A. Then do item B. Pencils down when you're finished. √
j. Item B. You started with the number 7,131,000. What did you write in the first box? (Signal.) *7.131.*
- Then you counted the places after the first digit to the end of the original number. How many places? (Signal.) *6.*
- So the exponent of 10 is 6.
- What does 7,131,000 equal? (Signal.) *$7.131 \times 10^6$.*

k. Work the rest of the problems in part 2. Pencils down when you're finished. (Observe students and give feedback.)

l. Check your work.
- I'll read the first value of each equation. You'll tell me what that value equals.
- Item C: 37,000. What does it equal? (Signal.) *3.7 × 10⁴.*
- Item D: 8040. What does it equal? (Signal.) *8.04 × 10³.*
- Item E: 70,180,000. What does it equal? (Signal.) *7.018 × 10⁷.*

═══ **Exercise 2** ═══
## ALGEBRA TRANSLATION
### Combination Symbols
━ **Textbook practice** ━

a. Open your textbook to lesson 92, part 1. √
- (Teacher reference:)

---
◆ If you had **at least** $103,
   you had  ≥  $103.

◆ If you had **a minimum of** 7 pets,
   you had  ≥  7 pets.

---

b. The box shows expressions that mean the same as **more than or equal to.**
- The first expression is **at least.**
- What's the expression? (Signal.) *At least.*
- **At least** is the same as **more than or equal to.**
- If you had at least $103, you would have **$103 or more than $103.**
- So you would use the symbol for **greater than or equal to.**

c. The next expression that means **more than or equal to** is **a minimum of.**
- What's the expression? (Signal.) *A minimum of.*
- If you had a minimum of 7 pets, you would have 7 pets or more than 7 pets.

d. Remember the new wording, **at least** and **a minimum of.** For both, you write **greater than or equal to.**

e. You're going to write a statement for each sentence.

f. Sentence A: 4J is at least 60.
- Can 4J be 60? (Signal.) *Yes.*
- Can 4J be more than 60? (Signal.) *Yes.*
- Write the statement. √
- (Write on the board:) [92:2A]

> **a.** $4j \geq 60$

- Here's what you should have.
- Everybody, read the statement. (Signal.) *4J is greater than or equal to 60.*

g. Sentence B: P is at most 54.
- Can P be 54? (Signal.) *Yes.*
- Can P be more than 54? (Signal.) *No.* P can be no more than 54.
- Write the statement. √
- (Write on the board:) [92:2B]

> **b.** $p \leq 54$

- Here's what you should have.
- Everybody, read the statement. (Signal.) *P is less than or equal to 54.*

h. Sentence C: The cost of the hats cannot exceed $12.
- Can the hats cost $12? (Signal.) *Yes.*
- Can the hats cost more than $12? (Signal.) *No.*
- Write the statement. √
- (Write on the board:) [92:2C]

> **c.** $h \leq 12$

- Here's what you should have.
- Everybody, read the statement. (Signal.) *H is less than or equal to 12.*

i. Sentence D: Shoes cost at least $45.
- Write the statement. √
- (Write on the board:) [92:2D]

> **d.** $s \geq 45$

- Here's what you should have.
- Everybody, read the statement. (Signal.) *S is greater than or equal to 45.*

j. Sentence E: Her savings did not exceed $5 a week.
- Write the statement. √

- (Write on the board:) [92:2E]

> **e.** $s \leq 5$

- Here's what you should have.
- Everybody, read the statement.
  (Signal.) *S is less than or equal to 5.*
k. Sentence F: Roberta had a minimum of 3 appointments.
- Write the statement. √
- (Write on the board:) [92:2F]

> **f.** $R \geq 3$

- Here's what you should have.
- Everybody, read the statement.
  (Signal.) *R is greater than or equal to 3.*

## Exercise 3
### SIMILAR TRIANGLES
### Corresponding Sides
#### — Textbook practice —

a. Find part 2. √
- (Teacher reference:)

- ◆ These are similar triangles, but they are not oriented the same way.
- ◆ You have to figure out the corresponding sides.
- ◆ First find the longest side of each triangle.
- ◆ Then find the shortest side of each triangle.

- These are similar triangles, but they are not oriented the same way.
- You have to figure out the corresponding sides.
- The simplest way to do that is to first find the longest side of each triangle. Then find the shortest side of each triangle.
b. Touch the longest side of each triangle. √
- You should be touching sides BD and HK.
c. Touch the shortest side of each triangle. √
- You should be touching sides DF and KJ.

d. Touch the last pair of corresponding sides. Those are the middle length sides. √
- You should be touching BF and JH.

#### — Textbook practice —
a. Find part 3 of your textbook. √
b. Item A. Touch the side with the question mark. √
- Is that the longest side, the shortest side, or the middle-length side? (Signal.) The shortest side.
- Touch the corresponding side in the other triangle. √
- Write the fraction for those sides. Then complete the equation with the pair of corresponding sides that have numbers in both triangles. Pencils down when you've done that much.
  (Observe students and give feedback.)
- (Write on the board:) [92:3A]

> **a.** $\dfrac{XZ}{3} = \dfrac{16}{5}$

- Here's the equation you should have:
  XZ over 3 = 16 over 5.
  You'll solve the equation later.
c. Item B. Touch the side with the question mark and the corresponding side. √
- Are you touching the longest sides, the middle-length sides, or the shortest sides? (Signal.) *The longest sides.*
- Skip 3 lines. Write the fraction for that pair of sides. Then complete the equation with the fraction for the corresponding sides that have numbers. Pencils down when you've done that much.
  (Observe students and give feedback.)
- (Write on the board:) [92:3B]

> **b.** $\dfrac{PQ}{13} = \dfrac{4}{5}$

- Here's the equation you should have:
  PQ/13 = 4/5.

*Sample Lesson 92* **41**

d. Write equations for items C and D. Don't solve the problems. Just write the equations. Leave space. Pencils down when you've done that much.
(Observe students and give feedback.)

e. Check your work.
Read the equation for each item.

• Item C. (Signal.) *RP/18 = 10/12.* (Accept *PR/18 = 10/12.*)

• Item D. (Signal.) *YZ/24 = 14/18.* (Accept *ZY/24 = 14/18.*)

f. Solve each problem. The unit name is centimeters. Pencils down when you're finished.
(Observe students and give feedback.)

g. Check your work.

• Item A. You figured out XZ. Everybody, what does XZ equal? (Signal.) *9 and 3/5 centimeters.*
Yes, 9 and 3/5 centimeters.

• Item B. You figured out PQ. Everybody, what does PQ equal? (Signal.) *10 and 2/5 centimeters.*
Yes, 10 and 2/5 centimeters.

• Item C. You figured out RP. Everybody, what does RP equal? (Signal.) *15 centimeters.*
Yes, 15 centimeters.

• Item D. You figured out YZ. Everybody, what does YZ equal? (Signal.) *18 and 2/3 centimeters.*
Yes, 18 and 2/3 centimeters.

——————— **Exercise 4** ———————

**PYTHAGORAS**

**Solve for *x* or *h***

━ **Textbook practice** ━

a. Find part 4. $\sqrt{}$

b. Each triangle shows the length of 2 sides. You'll figure out the length of the missing side. If the missing side is not the hypotenuse, call it X. Remember, if the missing side is opposite the right angle, it's the hypotenuse.

c. Triangle A. Does the problem give the hypotenuse? (Signal.) *Yes.*
So what do you solve for? (Signal.) *X.*

• Triangle B. Does the problem give the hypotenuse? (Signal.) *No.*
So what do you solve for? (Signal.) *H.*

• Triangle C. Does the problem give the hypotenuse? (Signal.) *Yes.*
So what do you solve for? (Signal.) X.

• Triangle D. Does the problem give the hypotenuse? (Signal.) *No.*
So what do you solve for? (Signal.) *H.*

• (Repeat step c until firm.)

d. Work the problems. Show the answer as a whole number or a square-root number. You don't need a calculator. Refer to the square-root table if you need to. It's shown on the inside back cover of your textbook. Pencils down when you're finished.
(Observe students and give feedback.)

e. Check your work.

f. Problem A. Did you figure out X or H? (Signal.) *X.*

• What does $X^2$ equal? (Signal.) *65.*

• What does X equal? (Signal.) $\sqrt{65}.$

g. Problem B. Did you figure out X or H? (Signal.) *H.*

• What does $H^2$ equal? (Signal.) *109.*

• What does H equal? (Signal.) $\sqrt{109}.$

h. Problem C. Did you figure out X or H? (Signal.) *X.*

• What does $X^2$ equal? (Signal.) *400.*

• What does X equal? (Signal.) *20.*

i. Problem D. Did you figure out X or H? (Signal.) *H.*

• What does $H^2$ equal? (Signal.) *50.*

• What does H equal? (Signal.) $\sqrt{50}.$

## Exercise 5

**PROBABILITY**

**— Textbook practice —**

a. Find part 5. √

• Some of these problems ask about the trials, and some ask about the object.

b. Problem A: A spinner has 24 equal-sized parts. If a person takes 96 trials and the spinner lands on a part that is green 16 times, how many parts are probably green?

• Does that problem ask about the object or the trials? (Signal.) *The object.*

c. Problem B: A deck has 100 cards, each with a different name on it. Lisa draws cards until she has drawn the name **Ed** 4 times. About how many trials does she take?

• Does that problem ask about the object or the trials? (Signal.) *The trials.*

d. Problem C: A game-show wheel has 18 prize spaces. The wheel is spun 190 times and lands on a prize space 90 times. How many spaces would you estimate are on the game-show wheel?

• Does that problem ask about the object or the trials? (Signal.) *The object.*

e. Problem D: A parking lot attendant collects keys to 112 cars. 42 of the cars are red. The parking lot attendant takes trials by randomly selecting keys to the cars. If the attendant takes 40 trials, on how many of those trials would you expect him to draw a key to a red car?

• Does that problem ask about the object or the trials? (Signal.) *The trials.*

f. Problem E. A deck contains 30 cards. Each card has a number or a picture on it. Lily takes 24 trials drawing a card, then replacing it in the deck. On 16 of those trials, the card has a picture on it. On how many cards in the deck would you expect there to be a picture?

• Does that problem ask about the object or the trials? (Signal.) *The object.*

g. Work problem A. Pencils down when you're finished.
(Observe students and give feedback.)

• (Write on the board:)                [92:5A]

$$a. \quad \frac{g}{24} = \frac{16}{96}$$

• Here's the equation you start with. You solved for the number of green parts the spinner probably has. What's the answer? (Signal.) *4 [green] parts.*

h. Work problem B. Pencils down when you're finished. (Observe students and give feedback.)

• Check your work. You solved for the total number of trials Lisa takes. What's the answer? (Signal.) *[About] 400 trials.*

i. Work problem C. Pencils down when you're finished.
(Observe students and give feedback.)

• Check your work. You solved for the total number of spaces on a game-show wheel. What's the answer? (Signal.) *38 spaces.*

j. Work problem D. Pencils down when you're finished.
(Observe students and give feedback.)

• Check your work. You solved for the number of trials you'd expect to draw a key to a red car. What's the answer? (Signal.) *15 trials.*

k. Work problem E. Pencils down when you're finished.
(Observe students and give feedback.)

• Check your work. You solved for the number of cards with a picture. What's the answer? (Signal.) *20 cards.*

## Exercise 6

**INDEPENDENT WORK**

> **Assign Independent Work: textbook parts 6–15 and workbook part 3.**

# Lesson 92

- Copy the digits before the final zeros.
- Write a decimal point after the first digit.
- Write the exponent for 10.

$$50{,}300 = \boxed{\phantom{xx}} \times 10^{\blacksquare}$$
$$5\ 03 \times 10^{\blacksquare}$$
$$5.03 \times 10^{\blacksquare}$$
$$5.03 \times 10^{4}$$

**Part 2**   **Complete each equation.**

a. $752{,}000 = \boxed{\phantom{xx}} \times 10^{\blacksquare}$

b. $7{,}131{,}000 = \boxed{\phantom{xx}} \times 10^{\blacksquare}$

c. $37{,}000 = \boxed{\phantom{xx}} \times 10^{\blacksquare}$

d. $8040 = \boxed{\phantom{xx}} \times 10^{\blacksquare}$

e. $70{,}180{,}000 = \boxed{\phantom{xx}} \times 10^{\blacksquare}$

## Independent Work

**Part 3**   **Plot the line for each equation.**

A   $y = -\dfrac{3}{4}x - 1$

B   $y = x - 4$

C   $y = -\dfrac{5}{4}x + 6$

D   $y = \dfrac{4}{3}x$

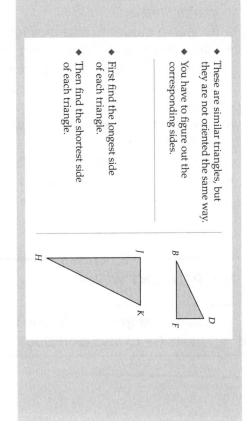

---

# Lesson 92

- If you had **at least** $103, you had ≥ $103.
- If you had **a minimum of 7 pets,** you had ≥ 7 pets.

**Part 2** ⟩ Similar Triangles

◆ Write a statement for each sentence.

a. 4j is at least 60.

b. p is at most 54.

c. The cost of the hats cannot exceed $12.

d. Shoes cost at least $45.

e. Her savings did not exceed $5 a week.

f. Roberta had a minimum of 3 appointments.

- These are similar triangles, but they are not oriented the same way.
- You have to figure out the corresponding sides.
- First find the longest side of each triangle.
- Then find the shortest side of each triangle.

## Part 3

For each pair of similar triangles, figure out the length of the side with a ?
All units are centimeters.

a.

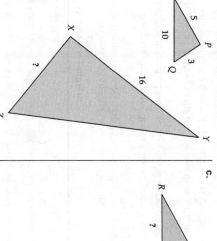

b.

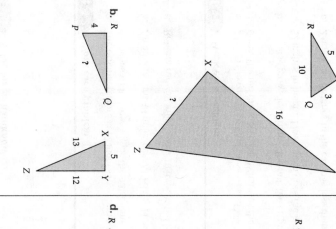

c.

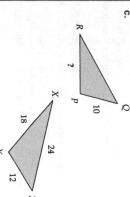

d.

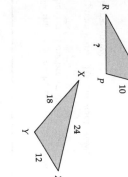

## Part 4

Figure out the missing value in each triangle.

a.  b.  c.  d.

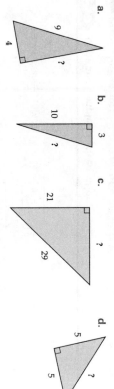
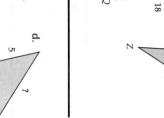

## Part 5

Work each item.

a. A spinner has 24 equal-sized parts. If a person takes 96 trials and the spinner lands on a part that is green 16 times, how many parts are probably green?

b. A deck has 100 cards, each with a different name on it. Lisa draws cards until she has drawn the name *Ed* 4 times. About how many trials does she take?

c. A game-show wheel has 18 prize spaces. The wheel is spun 190 times and lands on a prize space 90 times. How many spaces would you estimate are on the game-show wheel?

d. A parking lot attendant collects keys to 112 cars. 42 of the cars are red. The parking lot attendant takes trials by randomly selecting keys to the cars. If the attendant takes 40 trials, on how many of those trials would you expect him to draw a key to a red car?

e. A deck contains 30 cards. Each card has a number or a picture on it. Lily takes 24 trials drawing a card, then replacing it in the deck. On 16 of those trials, the card has a picture on it. On how many cards in the deck would you expect there to be a picture?

## Independent Work

## Part 6

Solve for each unknown.

a. $3(2x - 5) + 4 = 5(10 - x) + 5$

b. $-4(-5 - m) - 2m = 4$

## Part 7

Solve for both letters in each pair of equations.

a. $3g + 6 = 2q$

$9g - 9 = 3q$

b. $3r = 38 - 3v$

$-3r = -16 + 2v$

## Part 8

Copy and complete each equation.

a. $\sqrt{.81} = \blacksquare$

b. $\sqrt{\blacksquare} = 1.5$

c. $\sqrt{\blacksquare} = .4$

d. $\sqrt{\blacksquare} = 8$

## Part 9 ▶ Write an equation or inequality for each problem. Solve for the unknown.

**a.** 8 times $g$ is less than $\frac{1}{5}p$.

$$p = 280$$

**b.** 4 more than $\frac{5}{3}j$ is more than $2b$.

$$b = .25$$

## Part 10 ▶ Copy and complete each equation.

**a.** $3,971 \times 10^{\blacksquare} = 3971$

**b.** $.04 \times 10^{\blacksquare} = 400$

**c.** $29.3 \times 10^3 = \blacksquare$

## Part 11 ▶ Solve for each unknown.

**a.** $\frac{8}{5} - \frac{5}{5} = \frac{9}{r}$

**b.** $\frac{3}{m} = \frac{8}{2} \times \frac{5}{2}$

**c.** $\frac{P}{10} = \frac{7}{3} + \frac{11}{3}$

## Part 12 ▶ Work each problem.

**a.** The original picture is 12 inches high and 8 inches wide. The enlargement is 25 inches wide. How high is the enlargement?

**b.** An athlete runs 43 miles every 10 days. How far will she run in 4 days?

**c.** Roberta runs 52 miles every 10 days. What is her running rate per day?

## Part 13 ▶ Work each problem.

**a.** 2 times the cost of the red dress was $5 less than 3 times the cost of the brown dress. The red dress cost $56. What was the cost of the brown dress?

**b.** A horse grows less than $\frac{1}{4}$ the rate the elephant grows. The elephant grows 32 inches in height. How much does the horse grow?

**c.** 2 times the distance to the camp is less than $\frac{1}{2}$ the distance to the river. The distance to the river is 24 miles. What is the distance to the camp?

*Textbook Lesson 92*    383

## Part 14 ▶ Write the complete equation for each line.

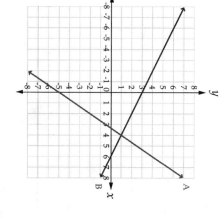

## Part 15 ▶ Work each problem.

**a.** If you double the weight of the rocks and subtract 90 pounds, you still have less than $\frac{1}{3}$ the weight of the truck. The weight of the rocks is 1045 pounds. What do you know about the weight of the truck?

**b.** 7 miles more than the distance to River City is less than $\frac{1}{2}$ the distance to Troutville. The distance to Troutville is 88 miles. What's the distance to River City?

**c.** 4 inches less than Vera's height is $\frac{2}{9}$ the height of the giraffe. Vera is 52 inches tall. What is the height of the giraffe?

384    *Lesson 92 Textbook*

**Part 1** ▷ Rewriting Large Numbers

- ◆ Copy the digits before the final zeros.
- ◆ Write a decimal point after the first digit.
- ◆ Write the exponent for 10.

$$50,300 = \blacksquare \times 10^{\blacksquare}$$
$$5\ 03 \times 10^{\blacksquare}$$
$$5.03 \times 10^{4}$$

**Part 2**  Complete each equation.

a. $752,000 = 7.52 \times 10^{5}$

b. $7,131,000 = 7.131 \times 10^{6}$

c. $37,000 = 3.7 \times 10^{4}$

d. $8040 = 8.04 \times 10^{3}$

e. $70,180,000 = 7.018 \times 10^{7}$

**Part 3** ▷ Independent Work

Plot the line for each equation.

A $\quad y = -\dfrac{3}{4}x - 1$

B $\quad y = x - 4$

C $\quad y = -\dfrac{5}{4}x + 6$

D $\quad y = \dfrac{4}{3}x$

Answer Key

Answer Key

---

**Lesson 92 Textbook**

**Part 1**

a. $4j \geq 60$
b. $p \leq 54$
c. $h \leq 12$
d. $s \geq 45$
e. $s \leq 5$
f. $R \geq 3$

**Part 3**

Letters can be in either order.

a. $(3)\ \dfrac{XZ}{3} = \dfrac{16}{5}\ (3)$
$XZ = \dfrac{48}{5}$

$9\frac{3}{5}$ cm

b. $(13)\ \dfrac{PQ}{13} = \dfrac{4}{5}\ (13)$
$PQ = \dfrac{52}{5}$

$10\frac{2}{5}$ cm

c. $(18)\ \dfrac{RP}{18} = \dfrac{10}{12}\ (18)$
$RP = 15$

$15$ cm

d. $(24)\ \dfrac{YZ}{24} = \dfrac{14}{18}\ (24)$
$YZ = \dfrac{56}{3}$

$18\frac{2}{3}$ cm

**Part 4**

a. $x^2 + y^2 = h^2$
$x^2 + 4^2 = 9^2$
$x^2 + 16 = 81$
$\phantom{x^2} -16 \quad -16$
$\sqrt{x^2} = \sqrt{65}$

$x = \sqrt{65}$

b. $3^2 + 10^2 = h^2$
$9 + 100 = h^2$
$\sqrt{109} = \sqrt{h^2}$

$\sqrt{109} = h$

c. $x^2 + 21^2 = 29^2$
$x^2 + 441 = 841$
$\phantom{x^2} -441 \quad -441$
$\sqrt{x^2} = \sqrt{400}$

$x = 20$

$$\begin{array}{r} 21 \\ \times\ 21 \\ \hline 21 \\ 420 \\ \hline 441 \end{array} \qquad \begin{array}{r} 29 \\ \times\ 29 \\ \hline 261 \\ 580 \\ \hline 841 \end{array}$$

d. $5^2 + 5^2 = h^2$
$25 + 25 = h^2$
$\sqrt{50} = \sqrt{h^2}$

$\sqrt{50} = h$

**Part 5**

object   trials

a. $(24)\ \dfrac{9}{24} = \dfrac{16}{96}\ (24)$
$g = 4$

$4$ green parts

b. $\dfrac{1}{100} = \dfrac{4}{t}$
$(4)\ \dfrac{100}{100} = \dfrac{t}{4}\ (4)$
$\dfrac{400}{1} = t$

about 400 trials

c. $(18)\ \dfrac{s}{18} = \dfrac{190}{48}\ (18)$
$s = 38$

$38$ spaces

d. $(40)\ \dfrac{42}{112} = \dfrac{r}{40}\ (40)$
$15 = r$

$15$ trials

e. $(30)\ \dfrac{P}{50} = \dfrac{16}{24}\ (30)$
$p = 20$

$20$ cards

## Lesson 92 Independent Work

### Part 6

a.
$$3(2x-5)+4 = 5(10-x)+5$$
$$6x-15+4 = 50-5x+5$$
$$6x-11 = 55-5x$$
$$\underline{+5x \qquad +5x}$$
$$11x-11 = 55$$
$$\underline{+11 \qquad +11}$$
$$11x = 66$$
$$\left(\tfrac{1}{11}\right)11x = 66\left(\tfrac{1}{11}\right)$$
$$\boxed{x = 6}$$

b.
$$-4(-5-m)-2m = 4$$
$$20+4m-2m = 4$$
$$20+2m = 4$$
$$\underline{-20 \qquad -20}$$
$$2m = -16$$
$$\left(\tfrac{1}{2}\right)2m = -16\left(\tfrac{1}{2}\right)$$
$$\boxed{m = -8}$$

### Part 7

a. $(-5)(3q+6) = 2q(-5)$
$$-9q-18 = -6q$$
$$9q \qquad 9q$$
$$-27 = -3q$$
$$27 = 3q$$
$$\left(\tfrac{1}{3}\right)27 = 3q\left(\tfrac{1}{3}\right)$$
$$\boxed{9 = q}$$

$$9q-9 = 3q$$
$$9q-9 = 3(9)$$
$$9q-9 = 27$$
$$\underline{+9 \qquad +9}$$
$$9q = 36$$
$$\left(\tfrac{1}{9}\right)9q = 36\left(\tfrac{1}{9}\right)$$
$$\boxed{g = 4}$$

b.
$$3r = 38-3v$$
$$\underline{-3v \qquad -3v}$$
$$0 = 22-v$$
$$\underline{+v \qquad +v}$$
$$\boxed{v = 22}$$

$$3r = 38-3v$$
$$3r = 38-3(22)$$
$$3r = 38-66$$
$$\left(\tfrac{1}{3}\right)3r = -28\left(\tfrac{1}{3}\right)$$
$$\boxed{r = -\tfrac{28}{3}}$$

**or**
$$-3r = -16+2v$$
$$-3r = -16+2(22)$$
$$-3r = -16+44$$
$$-3r = 28$$
$$\left(\tfrac{1}{3}\right)3r = -28\left(\tfrac{1}{3}\right)$$

### Part 8

a. $\sqrt{.81} = .9$
$$\begin{array}{r}.9\\ \times\, .9\\ \hline .81\end{array}$$

b. $\sqrt{2.25} = 1.5$
$$\begin{array}{r}1.5\\ \times\, 1.5\\ \hline 75\\ 150\\ \hline 2.25\end{array}$$

c. $\sqrt{.16} = .4$
$$\begin{array}{r}.4\\ \times\, .4\\ \hline .16\end{array}$$

d. $\sqrt{64} = 8$

### Part 9

a.
$$8g < \tfrac{1}{5}P$$
$$8g < \tfrac{1}{5}(280)$$
$$8g < 56$$
$$\left(\tfrac{1}{8}\right)8g < 56\left(\tfrac{1}{8}\right)$$
$$\boxed{g < 7}$$
$$\boxed{P = 280}$$

b.
$$5j+4 > 2b$$
$$5j+4 > 2(.25)$$
$$5j+4 > .5$$
$$\underline{-4 \qquad -4}$$
$$5j > -3.5$$
$$\left(\tfrac{1}{5}\right)5j > -3.5\left(\tfrac{1}{5}\right)$$
$$j > -\tfrac{10.5}{5}$$
$$\boxed{b = .25}$$

**or** $\boxed{j > -\tfrac{21}{10}}$ **or** $\boxed{j > -2.1}$

---

## Lesson 92 Part 10

a. $3.971 \times 10^{\boxed{3}} = 3971$

b. $.04 \times 10^{\boxed{4}} = 400$

c. $29.3 \times 10^3 = \boxed{29{,}300}$

### Part 11

a. $\dfrac{3}{5} = \dfrac{9}{r}$
$$(9)\tfrac{5}{3} = \tfrac{r}{9}(9)$$
$$\boxed{15 = r}$$

b. $\dfrac{3}{5} = \dfrac{10}{m}$
$$(3)\tfrac{m}{3} = \tfrac{10}{3}(3)$$
$$\boxed{m = \tfrac{5}{10}}$$

c. $(10)\dfrac{P}{10} = \dfrac{18}{3}(10)$
$$\boxed{P = 60}$$

### Part 12

a. $h = \left(\tfrac{h}{8}\right)w$
$$h = \left(\tfrac{12}{8}\right)25$$
$$h = \tfrac{75}{2}$$
$$\boxed{w = 25}$$
$$\boxed{37\tfrac{1}{2}\text{ inches}}$$

b. $m = \left(\tfrac{m}{d}\right)d$
$$m = \left(\tfrac{45}{10}\right)4$$
$$m = \tfrac{8g}{5}$$
$$\boxed{d = 4}$$
$$\boxed{5\tfrac{1}{5}\text{ miles per day}}$$

c. $\dfrac{m}{d} = \dfrac{m}{d}$
$$\tfrac{m}{d} = \tfrac{52}{10}$$
$$\boxed{e = 32}$$
$$\boxed{r = 24}$$

### Part 13

a.
$$2r = 3b-5$$
$$2(56) = 3b-5$$
$$112 = 3b-5$$
$$\underline{+5 \qquad +5}$$
$$117 = 3b$$
$$\left(\tfrac{1}{3}\right)117 = 3b\left(\tfrac{1}{3}\right)$$
$$39 = b$$
$$\boxed{r = 56}$$
$$\boxed{17\tfrac{1}{2}\text{ miles}}$$

b.
$$h < \tfrac{1}{4}e$$
$$h < \tfrac{1}{4}(52)$$
$$h < 8$$
$$\boxed{\text{less than 8 inches}}$$

c.
$$2c < \tfrac{1}{2}r$$
$$2c < \tfrac{1}{2}(24)$$
$$2c < 12\left(\tfrac{1}{2}\right)$$
$$c < 6$$
$$\boxed{\text{less than 6 miles}}$$

## Lesson 92 Part 14

A $y = \tfrac{3}{2}x - 5$

B $y = -\tfrac{1}{2}x + 3$

### Part 15

a.
$$2r-90 < \tfrac{1}{5}t$$
$$2(1045)-90 < \tfrac{1}{5}t$$
$$2090-90 < \tfrac{1}{5}t$$
$$2000 < \tfrac{1}{5}t$$
$$(5)2000 < \tfrac{1}{5}t(5)$$
$$6000 < t$$
$$\boxed{r = 1045}$$
$$\boxed{\text{more than 6000 pounds}}$$

b.
$$RC+7 < \tfrac{1}{2}T$$
$$RC+7 < \tfrac{1}{2}(88)$$
$$RC+7 < 44$$
$$\underline{-7 \qquad -7}$$
$$RC < 37$$
$$\boxed{T = 88}$$
$$\boxed{\text{less than 37 miles}}$$

c.
$$V-4 < \tfrac{2}{9}$$
$$52-4 < \tfrac{2}{9}$$
$$\left(\tfrac{9}{2}\right)48 < \tfrac{2}{9}\left(\tfrac{9}{2}\right)$$
$$216 < 9$$
$$\boxed{V = 52}$$
$$\boxed{216\text{ inches}}$$

*Answer Key*

# Tracks

Tracks are lesson segments that continue across more than one or two lessons. Each track teaches a particular topic. The development of content within a track occurs as a series of small steps, starting with the basic information and systematically adding details and extensions.

The scope and sequence chart on pages 2 and 3 shows the tracks in *Essentials for Algebra.*

As the scope and sequence chart shows, each lesson has more than one track—sometimes more than four. Throughout the program, skills are taught following the same general format. When a new skill or operation is introduced, it is highly structured. The same problem types will be repeated, but as the track progresses from one lesson to the next, you provide less structure until the only structure you provide is to tell students to work the problems.

At different times in the development of a skill or operation, the program introduces mixed sets of problems, those of the type that are being learned and those that have been learned earlier but that are similar to the type most recently introduced. At points in the program where the structure for working particular problem types is reduced and where mixed sets are introduced, students will have trouble and make mistakes. If you remember the sequence that was presented earlier, when the problems were more highly structured, you will be able to provide efficient corrections. You simply provide the structure and remind the students of the steps or details they are to attend to. Do not continue to present this structure any longer than is necessary. If students need constant corrections and structuring, they are misplaced in the program. The properly placed students will encounter some bumps in the road but should never be so bogged down that they need constant structuring. When problems are presented with little structure in the sequence, students should not proceed in the sequence until they are able to work problems of this type without additional structure.

Everything that is taught ultimately becomes part of the students' Independent Work. If students are not able to work specific problem types without additional structure, they will not be able to do the Independent Work successfully.

# Lessons 1–15

The tracks that occur in lessons 1–15 address basic tool skills that are used throughout the program. Students who place at Lesson 1 lack these skills. (Lessons 1–15 lend themselves well for use in a summer-school program because all the tracks in this lesson range are completed by lesson 15.) Students who place at Lesson 16 have most of these skills; however, you may find that you need to review some of the tracks in lessons 1–15 for students who start at lesson 16, particularly during the lesson range of 16–30.

There are six tracks for lessons 1–15:

Short Division

Decimal Rounding

Decimal Operations

Fraction Operations

Fraction, Decimal, Percent Equivalences

Geometry

Lessons 1–15 also provide brief reviews of word problems that require adding or subtracting dollar amounts; fraction properties and figures that represent fractions; and abbreviations for measurement units.

Note that the teaching provided in lessons 1–15 has less structure and is more abbreviated than the tracks that start in lesson 16. The reason is that students have been exposed more to the basic skills in lessons 1–15 than they have to the more sophisticated skills.

You should hold students to the same standards of performance throughout the program, however. Do not proceed until students achieve mastery on whatever you teach. If students are placed appropriately, they should be able to move through the program at a reasonable rate.

In general, they should be able to complete a lesson a day, and they should perform well enough that they achieve mastery without extensive re-teaching of what was taught earlier.

## Short Division (Lessons 3–6)

Typically, students who place at lesson 16 know short division but not long division. Students who place at lesson 1 often have a poor understanding of short division.

Within the track, students first work on problems that have a digit in the answer for every digit in the problem:

$$\begin{array}{r} 203 \\ 3\overline{)609} \end{array}$$

Next, they work simple problems in which there is no digit in the answer for the first digit of the problem:

$$\begin{array}{r} 50 \\ 7\overline{)350} \end{array}$$

Next, they work problems that require "internal carrying" of a remainder:

$$\begin{array}{r} 1\ 5\ 2 \\ 5\overline{)7_2 6_1 0} \end{array}$$

Finally, they work problems that have a remainder:

$$\begin{array}{r} 9\ 7\frac{1}{2} \\ 2\overline{)1\ 9_1 5} \end{array}$$

Note that the progression of types is cumulative.

The exercises are brief and provide practice on the type or types that have been introduced. Following is an example from lesson 4.

Students first work four problems that review the simplest problem type, which was presented in lesson 3.

In steps a–e, the teacher leads students through the procedure shown in the teaching box. In steps f and g, students work problems of the new type.

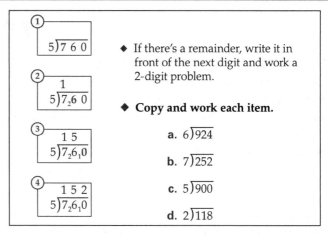

- The boxes show a new kind of problem.
- b. Everybody, read the problem in box ①. (Signal.) *760 ÷ 5.*
- Say the division problem for the first digit. (Signal.) *7 ÷ 5.*
- 5 goes into 7 **one** time, but there's a remainder.
- What's the remainder? (Signal.) *2.*
- c. Box ② shows that you write the remainder in front of the next digit. Then you work the problem: 26 divided by 5.
- 5 goes into 26 how many times? (Signal.) *5.*
- What's the remainder? (Signal.) *1.*
- d. Box ③ shows that you write the remainder in front of the last digit.
- Say the last division problem you work. (Signal.) *10 ÷ 5.*
- What's the answer? (Signal.) *2.*
- e. Box ④ shows the whole problem and the answer. Remember, if there's a remainder, write it in front of the next digit and work a 2-digit problem.
- f. Read problem A. (Signal.) *924 ÷ 6.*
- This is just like the problem shown in the boxes. Copy the problem and then stop. √
- (Write on the board:) [4:3B]

$$\textbf{a.} \quad 6\overline{)9\ 2\ 4}$$

- What's the first digit of 924? (Signal.) *9.*
- Say the problem for the first digit. (Signal.) *9 ÷ 6.*
- Write the answer above the 9 and show the remainder in front of the 2. √

- (Write to show:) [4:3C]

$$\begin{array}{r} 1\phantom{0} \\ 6\overline{)9\,{}_32\ 4} \end{array}$$ a.

- Here's what you should have.
- The problem for the next 2 digits is 32 divided by 6.
- Say the problem for those 2 digits. (Signal.) *32 ÷ 6.*
- Write the answer above the 2 and show the remainder in front of the 4. √
- (Write to show:) [4:3D]

$$\begin{array}{r} 1\ 5 \\ 6\overline{)9\,{}_32\,{}_24} \end{array}$$ a.

- Say the problem for the last 2 digits. (Signal.) *24 ÷ 6.*
- Write the answer above the 4. √
- (Write to show:) [4:3E]

$$\begin{array}{r} 1\ 5\ 4 \\ 6\overline{)9\,{}_32\,{}_24} \end{array}$$ a.

- Read the whole problem and the answer. (Signal.) *924 ÷ 6 = 154.*
g. Touch problem B. √
- Read it. (Signal.) *252 ÷ 7.*
- Say the problem for the first digit of 252. (Signal.) *2 ÷ 7.*
- How many times does 7 go into 2? (Signal.) *Zero.*
- Zero times. So you don't write anything in the answer.
- Say the problem for the first 2 digits of 252. (Signal.) *25 ÷ 7.*
- You can work that problem but it has a remainder.
- Copy problem B and work it. Pencils down when you're finished.
  (Observe students and give feedback.)

- (Write on the board:) [4:3F]

$$\begin{array}{r} 3\ 6 \\ 7\overline{)2\ 5\,{}_42} \end{array}$$ b.

- Here's what you should have.
- 25 divided by 7 is 3 with a remainder of 4.
- Say the 2-digit problem you worked next. (Signal.) *42 ÷ 7.*
- What's the answer? (Signal.) *6.*
- Say the whole division problem and the answer. (Signal.) *252 ÷ 7 = 36.*

*Teaching note:* Make sure you are in a position to monitor the performance of students in step f. Students are to work the problem a step at a time. The √ indicates that you observe whether they are working ahead. Note that the checking procedure early in the program takes a little longer than it does after students have learned the rules and practiced them. Do not permit students to work ahead of your directions. Remind them that they will first work a lot of problems a step at a time.

Note that the second problem provides far less structure than the first. If the students perform poorly, show the work on the board. Then direct students to copy the problem below the original on their lined paper. Make sure that the digits in the answer are directly above those in the problem.

The (Write on the board:) serves as a correction procedure for students who failed to get the right answer for each digit. If students tend to have trouble with problem A, structure problem B the same way the script directs the presentation of A. If students have trouble working C without help, they will probably have trouble on the next lesson because a new type will be introduced.

If students still require help on problem D, plan to repeat this exercise before presenting the first exercise of Lesson 5.

In lesson 6, the final short division type is introduced. These are problems with a remainder, which students express as a fraction. This convention prepares students to express answers to word problems as a mixed number and a unit name.

Here's the work for these problems as it is shown in the answer key for lesson 6.

a. $81\frac{2}{5}$  
5)407

c. $286\frac{1}{2}$  
2)5,7,3

e. $3002$  
3)9006

g. $60\frac{1}{6}$  
6)361

b. $205$  
9)1845

d. $826\frac{3}{4}$  
4)33,0,7

f. $1799$  
4)7,1,9,6

h. $1235$  
5)6,1,7,5

If students have serious problems, repeat the exercise the next day before proceeding to lesson 7.

## Decimal Rounding (Lessons 1–11)

Starting in lesson 1, students review reading decimal numbers with tenths, hundredths, and thousandths. In lessons 3–5, students round decimal values to the nearest whole number. The rule they learn is if the tenths digit is 5 or more, round up to the next whole number. Students next apply the rule to rounding decimal values to the tenths, hundredths, and thousandths place. The exercises in lessons 6–8 are highly structured and focus on rounding to the underlined digit. Below is the exercise from lesson 6.

---

You can round decimal values to tenths, hundredths, or any other decimal place.

When you round to **whole numbers,** you look at the **tenths** digit.

When you round to **tenths,** you look at the **hundredths** digit.

When you round to **hundredths,** you look at the **thousandths** digit.

---

b. You've learned to round decimal numbers to the nearest whole number.
- You can also round values to **tenths, hundredths,** or any other decimal place.

c. Listen:
- When you round to **whole numbers,** you look at the **tenths** digit.
- When you round to **tenths,** you look at the **hundredths** digit.
- When you round to **hundredths,** you look at the **thousandths** digit.

d. Once more:
- When you round to a **whole number,** which digit do you look at? (Signal.) *The tenths digit.*

- When you round to **tenths,** which digit do you look at? (Signal.) *The hundredths digit.*
- When you round to **hundredths,** which digit do you look at? (Signal.) *The thousandths digit.*
- (Repeat step d until firm.)

### — Textbook practice —

a. Find part 2. √
- (Teacher reference:)

---

| **a.** 25.7<u>0</u>9 | **c.** .08<u>6</u>2 | **e.** 9.05<u>4</u>9 |
| **b.** 110.6<u>8</u>52 | **d.** 1.2<u>8</u>14 | |

---

- You're going to round these values to the underlined place.

b. Value A: 25 and 709 thousandths. The tenths digit is underlined, so you round to tenths. That digit is underlined. So you look at the next digit.
- Touch the digit you'll look at. √
- What digit? (Signal.) *Zero.*

c. Value B: 110 point 6, 8, 5, 2. The hundredths digit is underlined. You round to hundredths.
- So you look at the next digit. Touch that digit. √
- What digit? (Signal.) *5.*

d. Value C: Point zero, 8, 6, 2. The tenths digit is underlined.
- So what do you round to? (Signal.) *Tenths.*
- What's the tenths digit? (Signal.) *Zero.*
- Remember, the digit you look at is not the underlined digit. Touch the digit you'll look at. √
- You should be touching the 8.

e. Value D: 1 point 2, 8, 1, 4. The thousandths digit is underlined.
- So what do you round to? (Signal.) *Thousandths.*
- What's the thousandths digit? (Signal.) *1.*
- Touch the digit you'll look at. √
- You should be touching the digit after thousandths.
- What digit? (Signal.) *4.*

f. Value E: 9 point zero, 5, 4, 9. The hundredths digit is underlined.

- So what do you round to? (Signal.) *Hundredths.*
- What's the hundredths digit? (Signal.) *5.*
- Touch the digit you'll look at. √
- What digit? (Signal.) *4.*

g. (Repeat any items that were not firm.)

h. For each item, you'll write the rounded decimal value. Remember, the underlined digit shows how many places will be in the rounded value.

i. Value A: 25 and 709 thousandths.
- What's the underlined digit? (Signal.) *7.*
- Tell me both numbers 7 **could** round to. (Signal.) *7 or 8.*
- Touch the digit you'll look at. √
- What digit? (Signal.) *Zero.*
- Is that 5 or more? (Signal.) *No.*
- When you write the rounded value, there will not be any digits after the tenths place. Write the rounded value. Pencils down when you're finished. √
- (Write on the board:)         [6:1A]

| | |
|---|---|
| **a.** | 25.7 |

- Here's what you should have: 25.7.

j. Value B. 110 point 6, 8, 5, 2.
- What's the underlined digit? (Signal.) *8.*
- Tell me both numbers 8 **could** round to. (Signal.) *8 or 9.*
- Touch the digit you'll look at. √
- What digit? (Signal.) *5.*
- Is that 5 or more? (Signal.) *Yes.*
- When you write the rounded value, there will not be any digits after the hundredths place. Write the rounded value. Pencils down when you're finished. √
- (Write to show:)         [6:1B]

| | |
|---|---|
| **a.** | 25.7 |
| **b.** | 110.69 |

- Here's what you should have. You rounded the underlined digit to 9.

k. Value C: Point zero, 8, 6, 2.
- What's the underlined digit? (Signal.) *Zero.*

- Tell me both numbers zero could round to. (Signal.) *Zero or 1.*
- Touch the digit you'll look at. √
- What digit? (Signal.) *8.*
- Is that 5 or more? (Signal.) *Yes.*
- Write the rounded value. Pencils down when you're finished. √
- (Write to show:)         [6:1C]

| | |
|---|---|
| **a.** | 25.7 |
| **b.** | 110.69 |
| **c.** | .1 |

- Here's what you should have. You rounded the underlined digit to 1.

l. Write the rounded values for items D and E. Pencils down when you're finished. (Observe students and give feedback.)

m. Check your work.
- (Write to show:)         [6:1D]

| | |
|---|---|
| **c.** | .1 |
| **d.** | 1.281 |
| **e.** | 9.05 |

- Here are the rounded values you should have.

*Teaching note:* In step d, the exercise tests students on the basic rule about where to find the information that guides the rounding. Students are to respond orally. Expect to repeat step d several times. The rule that students learn is not easy for them to apply. It becomes easier if they have a solid idea of looking at the next digit.

Students do textbook practice next. They apply the rule in steps d–f. Note that this is another procedure you direct a step at a time.

Step g directs you to repeat any value the students got wrong.

Next, students work the problems they have verbally analyzed. The first three problems are structured and checked before students work the remaining problems from global directions. (Write the rounded values for items D and E.

Pencils down when you're finished.) When you observe students who made a mistake, tell them they didn't follow the rule about rounding. Touch the underlined digit and ask:

- Is the underlined digit a whole number, tenths, hundredths, or thousandths?

- Touch the digit you'll look at. √

- Will you round up the underlined digit or will it stay the same?

Again, make sure students follow directions, working both problems and putting their pencils down when they're finished.

The last exercises in the track (lessons 9–11) do not show underlined digits. Directions indicate how each value is to be rounded. Each value is rounded to tenths, hundredths, or thousandths. In the following lessons, students practice rounding values to different places as part of their Independent Work.

## Decimal Operations (Lessons 4–15)

Students review basic decimal procedures: aligning decimal points for adding and subtracting and counting the sum of the decimal places for multiplication to determine the number of decimal places in the answer.

In lessons 4–6, students practice working decimal add-subtract problems in columns. They review the rule about adding zeros to the end of numbers that have too few decimal places. Here is part of an exercise from Lesson 5 that applies the procedure to subtraction.

| $5.4 - 1.02$ | ◆ If the number you subtract has more decimal places than the top number, you need to **add zeros**. | $33 - 5.02$ |
|---|---|---|
| $5.40$ | | $33.00$ |
| $-\ 1.02$ | | $-\ 5.02$ |
| $4.38$ | | |

b. The box shows the problem 5.4 minus 1.02.

- You can see that problem written in a column.

- You can see a zero written in the hundredths place for 5 and 4/10. You need that zero so you can subtract.

- Now you write the decimal point in the answer and work the problem like any other subtraction problem. The answer is 4.38.

c. Remember, if the number you subtract has more decimal places than the top number, you need to add zeros.

d. Read the next problem. (Signal.) *33 – 5.02.*

- How many decimal places does 5.02 have? (Signal.) *2.*

- That's how many places 33 needs to have. So you rewrite 33 as 33 and zero hundredths. Then you work the problem.

*Teaching note:* The program makes no attempt to teach subtraction of the type shown in the problem above. Students are assumed to have knowledge of the mechanics of regrouping.

As a general rule, the program provides sufficient practice to demonstrate how to perform different operations and to provide feedback on possible errors.

Do not labor explanations provided by the box because the procedure will become evident when students work the first problem. This is the point at which you would bring students to mastery.

Multiplication problems with decimal values are introduced in lesson 9, after students have worked on add-subtract problems for several days. For the first five days of the multiplication sequence, students do not work entire problems but simply add the decimal point to the answers of problems that are already worked.

Here's the exercise from lesson 10, the second day of the sequence.

| a. | 20.03 | c. | 5.3 | e. | 21.7 |
|---|---|---|---|---|---|
| | $\times\ \ .11$ | | $\times\ 85$ | | $\times\ 4.6$ |
| | 2003 | | 265 | | 1302 |
| | 20030 | | $+\ 4240$ | | $+\ 8680$ |
| | 22033 | | 4505 | | 9982 |
| b. | 5.64 | d. | .256 | | |
| | $\times\ \ .4$ | | $\times\ 1.01$ | | |
| | 2256 | | 256 | | |
| | | | $+\ 25600$ | | |
| | | | 25856 | | |

- These are multiplication problems that involve decimal values. You'll copy the digits for each answer and put the decimal points in the right place.
- Remember, you add up the decimal places in the problem, and you show the same number of decimal places in the answer.
b. Problem A. How many decimal places in 20 and 3 hundredths? (Signal.) *2.*
- How many decimal places in 11 hundredths? (Signal.) *2.*
- So how many decimal places will be in the answer? (Signal.) *4.*
- Copy the answer for problem A and show the decimal point. Remember, count places from the end of the number, not the beginning. Pencils down when you're finished.
  **(Observe students and give feedback.)**
- (Write on the board:)                    [10:5A]

> **a.    2.2033**

- Here's what you should have. There are 4 places after the decimal point.
c. Write answers for the rest of the problems in part 5. Show the correct number of decimal places. Pencils down when you're finished.
  **(Observe students and give feedback.)**
d. Check your work.
- (Write to show:)                        [10:5B]

> **a.    2.2033      d.    .25856**
> **b.    2.256       e.    99.82**
> **c.    450.5**

- Here's what you should have for each decimal answer.

In lessons 14 and 15, students work multiplication problems and place the decimal points.

Also in lessons 14 and 15, a mixed set of problems appears in the Independent Work. Pay particular attention to how students perform on this part of lesson 14 (part 6). If they do poorly, give them feedback and repeat that part before presenting lesson 15. If problems persist in lesson 16, follow the same procedure. If students are at mastery after lesson 15, they will make fewer mistakes when they apply the decimal placement rules to problem types introduced later.

## Fraction Operations (Lessons 1–6)

In lessons 1 and 2, students review and apply the procedure that when they add or subtract, the denominators must be the same in all fractions (including the fraction in the answer).

In lessons 3 and 4, they review and apply the procedure that when they multiply, the denominators need not be the same. To work the problem, students operate on the numbers in the numerator and those in the denominator.

In lessons 5 and 6, students work mixed sets of problems, some involving addition or subtraction of fractions, others involving multiplication.

Here's the exercise for introducing the mixed set in lesson 6.

| | |
|---|---|
| a. $\dfrac{3}{v} - \dfrac{m}{7}$ = ■ | e. $\dfrac{7}{5} + \dfrac{4}{5}$ = ■ |
| b. $\dfrac{3}{v}\left(\dfrac{m}{7}\right)$ = ■ | f. $\dfrac{7}{5} \times \dfrac{3}{5}$ = ■ |
| c. $\dfrac{9}{11} - \dfrac{7}{11}$ = ■ | g. $\dfrac{12}{a} - \dfrac{9}{a}$ = ■ |
| d. $\dfrac{6}{5} + \dfrac{6}{9}$ = ■ | h. $\dfrac{5}{r} + \dfrac{9}{4r}$ = ■ |

- Some of these problems multiply fractions; some add; some subtract.
- You can work all the multiplication problems the way they are written.
- You can't work all the addition or subtraction problems the way they are written.
b. Write the letters of all problems you can work the way they are written. Pencils down when you've done that much.
  **(Observe students and give feedback.)**
c. Check your work.
- (Write on the board:)                    [6:4A]

> **b.    c.    e.    f.    g.**

- Here are the letters of all the problems you can work the way they are written: B, C, E, F, G.

d. Copy these problems and work them. Pencils down when you're finished. (Observe students and give feedback.)

e. Check your work.
- You'll read the equation for each problem.
f. Equation B. (Signal.) *3/V (M/7) = 3M/7V.*
- Equation C. (Signal.) *9/11 − 7/11 = 2/11.*
- Equation E. (Signal.) *7/5 + 4/5 = 11/5.*
- Equation F. (Signal.) *7/5 × 3/5 = 21/25.*
- Equation G. (Signal.) *12/A − 9/A = 3/A.*

> ***Teaching note:*** Students were introduced to parentheses in lesson 4. Monitor the students as they work the problems (step d of the exercise) and remind those who make mistakes of the rules: For multiplication problems, multiply in the denominator. For addition or subtraction problems, copy the denominator.

Students are not taught to work add-subtract problems that have fractions with unlike denominators. One reason for this omission is that it is not often required for various exit exams. Furthermore, it is not needed for any of the sophisticated applications presented in *Essentials.* Finally, teaching this skill requires a major effort. The typical student who places at lesson 1 is weak on all fraction concepts, including problems that have the same denominator.

## Fraction Equivalences (Lessons 1–15)

This track is critical because it introduces skills and discriminations students will use throughout the program (and throughout their studies in math). Furthermore, students are typically deficient in their understanding of fractions and fraction equivalences. Students typically understand the procedures for working with fractions that are less than one but don't understand that fractions have a much greater range. The purpose of this track is to induce the general understanding that any whole number, mixed number, fraction, or decimal value may be expressed as an equivalent fraction. In lessons 1–15, students learn (or review) the logic of fraction equivalences.

In lessons 1–3, students learn how to convert a fraction into a whole number or mixed number by division. Next, students learn that a whole number equals an indefinitely large number of equivalent fractions (lessons 5–9). Lessons 8–15 introduce equivalent-fraction problems of the form

$$\frac{3}{5} = \frac{\blacksquare}{15}$$

In lessons 10–13, students convert mixed numbers into fractions. In lessons 12–15, students determine whether pairs of fractions are equivalent. Also, during this lesson range, students express equivalence between fractions, decimal values, and percents.

### FRACTIONS AS DIVISION

In lessons 1–3, students learn to read fractions as division problems and convert fractions that are more than one into a whole number or mixed number by dividing.

Here's part of the introduction from lesson 1:

> a. $\frac{17}{5}$

- If a fraction is more than 1, here's how you figure out if it's a mixed number or a whole number. You say the fraction as a division problem. Then you work that problem.
- Remember, say the fraction as a division problem and then work it.
b. Fraction A is 17 fifths.
- I'll say 17 fifths as a division problem: 17 divided by 5.
- Listen again: 17 divided by 5.
- Your turn: Say 17 fifths as a division problem. (Signal.) *17 divided by 5.*
c. Fraction B: 31 sixths.
- Say the division problem. (Signal.) *31 divided by 6.*
d. Fraction C: 81 eightieths.
- Say the division problem. (Signal.) *81 divided by 80.*
e. Fraction D: 50 tenths.
- Say the division problem. (Signal.) *50 divided by 10.*
- (Repeat steps c–e until firm.)

f. (Write on the board:)  [1:4B]

$$5\overline{)17}$$
a.

- Here's fraction A written as a division problem.
- Everybody, read the problem. **(Signal.)** *17 divided by 5.*
- Now we'll work the problem.
- How many times does 5 go into 17? **(Signal.)** *3.*
- (Write to show:)  [1:4C]

$$5\overline{)17}^{\,3}$$
a.

- So the whole number is 3.
- There's a remainder. 5 × 3 is 15. Raise your hand when you know the remainder. √
- What's the remainder? **(Signal.)** *2.* Yes, the remainder is 2.
- So you write 2 over the number you divide by. That's 5. So the remainder is 2 fifths.
- (Write to show:)  [1:4D]

$$5\overline{)17}^{\,3\frac{2}{5}}$$
a.

- 17 fifths equals 3 and 2 fifths.
- (Write to show:)  [1:4E]

$$5\overline{)17}^{\,3\frac{2}{5}}$$
a.

$$\boxed{\frac{17}{5} = 3\frac{2}{5}}$$

- Copy the work for fraction A. √

*Teaching note:* Students often do not have a solid notion of the relationship between fractions and division. Make sure they are reliable in responding to your directions in steps c, d, and e.

Students express the remainder as a fraction. They will follow this convention for any division problems that are not to be solved as decimal problems.

In lesson 5, students multiply a whole number by one, but they replace the 1 with a fraction equal to one.

$$5 \times 1 = 5$$
$$5 \times \frac{8}{8} = \frac{40}{8} = 5$$

The exercise demonstrates that the fraction 40/8 must equal five because five is multiplied by one, and 5 × 1 = 5.

In lesson 7, students complete strings of fractions that are equivalent to whole numbers. They multiply by a series of fractions that equal one.

a. $5 = \dfrac{\ }{\ } = \dfrac{\ }{\ } = \dfrac{\ }{\ } = \dfrac{\ }{\ }$

b. $3 = \dfrac{\ }{\ } = \dfrac{\ }{\ } = \dfrac{\ }{\ } = \dfrac{\ }{\ }$

b. For item A, you're going to write a whole set of fractions that equal 5.
- Remember, 5 × 1 is 5, so 5 times any fraction that equals 1 is 5.
- Listen: For the first fraction, we'll work the problem 5 × 2/2 and write the answer in the first blank.
- (Write on the board:)  [7:1A]

a. $5 = \underline{\ \ }$

- What's 5 × 2? **(Signal.)** *10.*
- So the numerator is 10.
- (Write to show:)  [7:1B]

a. $5 = \underline{10}$

- We're multiplying 5 by 2/2.
- What's the denominator? **(Signal.)** *2.*

- (Write to show:)  [7:1C]

$$a. \quad 5 = \frac{10}{2}$$

- Everybody, what fraction does 5 × 2/2 equal? (Signal.) *10/2.*
- Write that fraction in the first fraction box. √
- c. For the next fraction, you'll multiply 5 by 3/3. Do it and write the fraction in the next box. √
- Everybody, what fraction does 5 × 3/3 equal? (Signal.) *15/3.*
- (Write to show:)  [7:1D]

$$a. \quad 5 = \frac{10}{2} = \boxed{\frac{15}{3}} =$$

- d. Write the next fraction. Work the problem 5 × 4/4. √
- Everybody, what does 5 × 4/4 equal? (Signal.) *20/4.*
- (Write to show:)  [7:1E]

$$a. \quad 5 = \frac{10}{2} = \frac{15}{3} = \boxed{\frac{20}{4}} =$$

- e. For the last fraction, work the problem 5 × 5/5. Pencils down when you're finished. √
- Everybody, what does 5 × 5/5 equal? (Signal.) *25/5.*
- (Write to show:)  [7:1F]

$$a. \quad 5 = \frac{10}{2} = \frac{15}{3} = \frac{20}{4} = \boxed{\frac{25}{5}}$$

- f. Everybody, read this whole equation. (Signal.) *5 = 10/2 = 15/3 = 20/4 = 25/5.*
- Listen: All of these fractions are equal. They all equal 5 because the numerator is 5 times greater than the denominator.

g. Item B. You'll multiply 3 by these fractions: 2/2, 3/3, 4/4, and 5/5.
h. Here's the first problem you'll work: 3 × 2/2.
- Say that problem. (Signal.) *3 × 2/2.*
i. Say the next problem you'll work. (Signal.) *3 × 3/3.*
- Say the next problem you'll work. (Signal.) *3 × 4/4.*
- Say the next problem you'll work. (Signal.) *3 × 5/5.*
- (Repeat steps g, h, and i until firm.)
j. Work the problems and fill in the missing fractions. Pencils down when you're finished.
   (Observe students and give feedback.)
k. Check your work.
- (Write on the board:)  [7:1G]

$$b. \quad 3 = \frac{6}{2} = \frac{9}{3} = \frac{12}{4} = \frac{15}{5}$$

- Here's what you should have.
- Everybody, read the whole equation. (Signal.) *3 = 6/2 = 9/3 = 12/4 = 15/5.*
l. Remember, all those fractions equal 3 because the numerator is 3 times greater than the denominator.

*Teaching note:* A good idea is to rehearse before presenting an exercise like this one. There are two reasons: First, the correction you would provide for mistakes involves reminding students of the relationship between the fraction and the whole number. Five times any fraction that equals one must equal five.

Second, rehearsing the exercise enables you to present the exercise with good pacing. Do not labor parts. If students make mistakes, tell them the answer. If they make more than one or two mistakes, start over and repeat the part of the exercise that students missed.

The key steps of the exercise that students must master before they proceed are g, h, and i. If students produce weak responses on these steps, repeat the sequence before directing them to work problem B.

Point out that one way to check the fractions is to note progression of denominators is counting by ones—2, 3, 4, 5—and the progression of numerators is counting by the whole number. For 5, the numerators are 10, 15, 20, 25. For 3, the numerators are 6, 9, 12, 15.

The terminal exercises in the track require students to identify which fractions in a string are not equivalent. The procedure they follow is to start with the first fraction in the string and then test whether the comparison fraction is obtained by multiplying by a fraction equal to one. Here is the exercise from lesson 13.

$$\frac{2}{10} = \frac{20}{80} = \frac{12}{60} = \frac{18}{90} = \frac{10}{40} = \frac{10}{50}$$

b. Touch the first fraction. √
- Everybody, what fraction? (Signal.) *2/10.*
- All the fractions are supposed to equal 2/10, but some of them are wrong. Remember, if you multiply 2/10 by a fraction equal to 1, the fractions are equivalent. If you don't multiply by a fraction equal to 1, the fractions are not equivalent.
c. The first fraction with a box under it is 20/80. That fraction is supposed to equal 2/10.
- Say the problem for the numerator. (Signal.) *2 × some value = 20.*
- What's the answer? (Signal.) *10.*
- Say the problem for the denominator. (Signal.) *10 × some value = 80.*
- What's the answer? (Signal.) *8.*
- So the fraction you multiply 2/10 by is 10/8.
- Does that fraction equal 1? (Signal.) *No.*
- So does 2/10 = 20/80? (Signal.) *No.*
- Write 10/8 in the box below 20/80. Then change the sign so it says 2/10 is not equal to 20/80. √

- (Write on the board:) [13:1A]

$$\frac{2}{10} \neq \frac{20}{80}$$

$$\boxed{\frac{10}{8}}$$

- Here's what you should have.
d. The next fraction is 12/60.
- Say the problem for the numerator. (Signal.) *2 × some value = 12.*
- Say the problem for the denominator. (Signal.) *10 × some value = 60.*
- Write the fraction you multiply 2/10 by in the box below 12/60. Change the sign if 12/60 does not equal 2/10. Pencils down when you're finished.
  (Observe students and give feedback.)
- Everybody, what fraction do you multiply 2/10 by to get 12/60? (Signal.) *6/6.*
- So are the fractions equivalent? (Signal.) *Yes.*
e. Work the rest of the fractions. Remember to say the problem for finding the missing number for the numerator and for the denominator. Then write the fractions and change the sign for the fractions that are not equivalent. Pencils down when you're finished.
  (Observe students and give feedback.)
f. Check your work.
g. The next fraction is 18/90.
- What do you multiply 2/10 by to get 18/90? (Signal.) *9/9.*
- So are 2/10 and 18/90 equivalent? (Signal.) *Yes.*
h. The next fraction is 10/40.
- What do you multiply 2/10 by to get 10/40? (Signal.) *5/4.*
- So are 2/10 and 10/40 equivalent? (Signal.) *No.*
- So you change the sign to say 2/10 ≠ 10/40.

- (Write on the board:)    [13:1B]

$$\neq \frac{10}{40}$$

i. The next fraction is 10/50.
- What do you multiply 2/10 by to get 10/50? (Signal.) *5/5.*
- So are 2/10 and 10/50 equivalent? (Signal.) *Yes.*

*Teaching note:* In earlier exercises, students have learned to say the problem for figuring out a missing factor. For example:

$$\frac{5}{6} \times \frac{\blacksquare}{\blacksquare} = \frac{15}{18}$$

However, the exercise in lesson 13 presents a different context. Nothing is written to guide the students in saying the problem for numerator or denominator. In steps c and d, students say the problem for figuring out the missing value: "Two times some value equals 20." Note that they are not to say "2 times 10 equals 20." They are to state the problem, not the fact. Be very strict about this convention. Students who learn to follow directions precisely will have less trouble later in the program.

In step c, some students may make the mistake of saying, "Two times 10. . ." for the numerator.

Correct by telling the correct answer, repeating the questions for both the numerator and denominator until students are firm on both. Then direct them to write the fraction, 10/8.

Students respond to fractions that are not equivalent by changing the equal sign into a not-equal sign. This sign was introduced in lesson 12.

## Fraction, Decimal, Percent Equivalences (Lessons 12–15)

In lesson 12, students complete a table that shows equivalent fractions and percents. They learn a percent number is a hundredths number. They then complete a table that shows fractions in one column and corresponding percents in the other column.

|   | Fraction | Percent |
|---|---|---|
| a. | $\frac{285}{100}$ | |
| b. | | 16% |
| c. | $\frac{7}{100}$ | |
| d. | $\frac{100}{100}$ | |
| e. | | 325% |

In lessons 13–15, students complete tables to show corresponding fractions, percents, and hundredths decimal values.

|   | Decimal | Fraction | Percent |
|---|---|---|---|
| | 2.06 | $\frac{206}{100}$ | 206% |
| a. | | $\frac{186}{100}$ | |
| b. | | $\frac{80}{100}$ | |
| c. | | $\frac{7}{100}$ | |
| d. | | $\frac{15}{100}$ | |
| e. | | $\frac{258}{100}$ | |

## Geometry (Lessons 1–15)

Perimeter is reviewed in lessons 1 and 2; area of rectangles in lessons 3–5; perimeter and area of rectangles in lessons 7–9; area of both triangles and rectangles in lessons 10–12; and mixed sets of problems involving triangles and rectangles are presented in lessons 14 and 15.

The rule students follow for finding the perimeter of a figure is, to find the perimeter, add the length of each side.

The formula they learn for area of rectangles is area equals base times height ($A = b \times h$).

In lessons 7–9, students find both the perimeter and the area of rectangles. They review the discrimination of unit names for area and perimeter in lesson 7.

Here's the exercise:

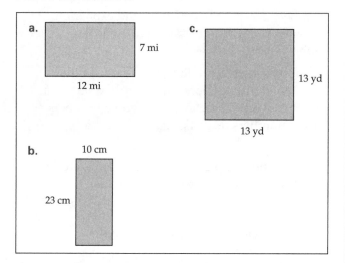

a. 7 mi, 12 mi

b. 10 cm, 23 cm

c. 13 yd, 13 yd

- You'll find the area and perimeter of each rectangle.

b. Listen: If you find the **area** of a rectangle that shows inches, what's the unit name for the area? (Signal.) *Square inches.*
- Listen: If you find the **perimeter** of a rectangle that shows inches, what's the unit name for the perimeter? (Signal.) *Inches.*
- If you find the **perimeter** of a rectangle that shows meters, what's the unit name for the perimeter? (Signal.) *Meters.*
- If you find the **area** of a rectangle that shows miles, what's the unit name for the area? (Signal.) *Square miles.*
- (Repeat step b until firm.)

c. Find the area and the perimeter for rectangle A. Pencils down when you're finished.
  (Observe students and give feedback.)
- Rectangle A. Everybody, what's the **area?** (Signal.) *84 square miles.*
- What's the perimeter? (Signal.) *38 miles.*

d. Find the area and the perimeter for rectangle B. Pencils down when you're finished.
  (Observe students and give feedback.)
- Rectangle B. Everybody, what's the **area?** (Signal.) *230 square centimeters.*
- What's the perimeter? (Signal.) *66 centimeters.*

e. Find the area and the perimeter for rectangle C. Pencils down when you're finished.
  (Observe students and give feedback.)
- Rectangle C. What's the **area?** (Signal.) *169 square yards.*
- What's the perimeter? (Signal.) *52 yards.*

In lesson 10, students apply the formula for finding the area of triangles:

$$A = \frac{1}{2}(b \times h)$$

They find the area of triangles that are shown as 1/2 of a rectangle with the same base and height.

Here's part of the exercise from lesson 10:

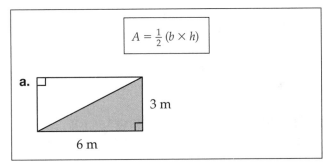

$A = \frac{1}{2}(b \times h)$

a.

3 m

6 m

f. You'll find the area of a triangle by multiplying 1/2 base times height.
• Listen: Say the equation for the area of a triangle. (Signal.) *Area = 1/2 base × height.*
• Write that equation with letters. Pencils down when you've done that much. √
• (Write on the board:)      [10:2A]

a.     $A = \frac{1}{2}(b \times h)$

• Here's what you should have.
• I'll show you how to work this problem.
• You put in the numbers for base and height.
• (Write to show:)      [10:2B]

a.     $A = \frac{1}{2}(b \times h)$

$A = \frac{1}{2}(6 \times 3)$

• Next you figure out the value inside the parentheses. What's 6 times 3? (Signal.) *18.*
• So you write A = 1/2 (18).

• (Write to show:)      [10:2C]

a.     $A = \frac{1}{2}(b \times h)$

$A = \frac{1}{2}(6 \times 3)$

$A = \frac{1}{2}(18)$

• Now you multiply. You get 18/2.
• Say the division problem for 18/2. (Signal.) *18 ÷ 2.*
• What's the answer? (Signal.) *9.*
• So the area of the triangle is 9 square meters.
• (Write to show:)      [10:2D]

a.     $A = \frac{1}{2}(b \times h)$

$A = \frac{1}{2}(6 \times 3)$

$A = \frac{1}{2}(18)$

9 sq m

• Copy the work for problem A. √

**Teaching note:** Students follow a process that they use throughout the program: writing the equation with letters, rewriting the equation with numbers, solving the equation, and expressing the answer as a number and a unit name (abbreviation).

Make sure students follow these steps. In the last step, students apply what they have learned about converting fractions into whole numbers or mixed numbers by working a division problem.

Following the completion of lessons 1–15, students take Mastery Test 1B. If they pass the test, they continue in the program. If they do not pass, provide the necessary remedies and then retest them. (See In-Program Mastery Tests, pages 25–28).

The foundation provided by lessons 1–15 gives students practice with the basic templates of solving problems they will follow in the rest of the program, not by writing a few numbers and trying to intuit an answer but by starting with a letter equation, writing a parallel equation with numbers, and solving the problem a step at a time.

# Lessons 16–118

## Rate Equations (Lessons 16–116)

This is an elaborate track that addresses many subtypes of problems involving rate, ratios, and proportion.

*Essentials for Algebra* expresses them as rate equations of the following form:

$$A\left(\frac{B}{A}\right) = B.$$

First students learn how to solve a numerical problem of the same form. Much of the track focuses on deciphering word problems. The simplest rate problems are of the following form:

> **There are 6 cans in each box. If there are 162 cans, how many boxes are there?**

The more difficult types require students to extract information from a table or work with a pair of equations needed to work the problem. Students also work problems with mixed units (feet and inches, pounds and tons).

Note that the track includes many problem types that are traditionally expressed as ratio equations. The main value in expressing these problems as rate equations rather than ratio equations is that we are able to provide students with a fairly ironclad procedure for writing the letter equation for the problem, and the computation requires fewer steps.

The procedure they learn is to identify the unit named in the question, the unit that answers the question the problem asks. Students write a letter equation based on that unit name.

For instance, if the problem asks how many students are in six groups, the student writes a letter equation in which *students* is the term after the equal sign.

$$g\left(\frac{s}{g}\right) = s$$

If the problem asks how many miles a train went in four hours, students write an equation in which *miles* is the term after the equal sign.

$$h\left(\frac{m}{h}\right) = m$$

If the problem asks how many hours it takes to go 50 miles, students write an equation in which *hours* is the term after the equal sign.

$$m\left(\frac{h}{m}\right) = h$$

If the equation is set up according to this convention, students will always solve for the letter after the equal sign.

To learn this application, students first learn how to identify the unit the problem asks about, how to write an equation based on that unit, and how to recognize parts of the problem that suggest it calls for a rate equation.

The early problems clearly name the units. Later problems ask How long did it take? How much did it weigh? How long is it? and so forth. Students are next introduced to ratio language and identify parts of word problems that refer to the *related units*—5 miles every 2 days; 6 papers an hour.

If the problem asks about rate or the pair of related units, students write a *simple* equation. For the question "What's the boat's average rate in miles per hour?" students write a simple equation.

$$\frac{m}{h} = \frac{m}{h}$$

Throughout the track, each new problem type that is taught is discriminated from all other types that have been taught but that might be confused with the newly taught type. For example, after students learn to solve for rate, they work mixed sets of problems, some of which ask about rate, as well as the various types that have been taught earlier in the sequence. Students also work sets that include a mix of word problems that call for rate equations and word problems that call for algebra translation. (See page 89.)

The preskills for the rate track begin in lesson 16. Students first learn that any simple multiplication equation of the form 2 ( ) = 6 may be written with only two values, 2 and 6. The missing fraction has the number that is after the equal sign as the numerator:

$$2\left(\frac{6}{2}\right) = 6$$

Here's the introduction from lesson 16.

| $4\,(2) = 8$ $4\left(\frac{8}{4}\right) = 8$ $\frac{32}{4} = 8$ | ◆ The number **after** the equal sign is always the **numerator** of the fraction. ◆ The **first number** in the problem is always the **denominator** of the fraction. |  $5\left(\frac{\blacksquare}{\blacksquare}\right) = 2$ |
|---|---|---|

b. These equations show that one of the numbers you multiply by can be shown as a fraction.

c. Touch the top equation in the first box. √
• Everybody, read it. (Signal.) *4 times 2 equals 8.*

d. Touch the next equation. √
• That equation shows that you can rewrite **2** as a fraction made up of the other numbers in the problem: 4 and 8.
• What's the fraction? (Signal.) *8 fourths.* Yes, 8 fourths.

e. 8 fourths equals 2. So if you multiply 4 by 8 fourths, you get the same answer you do when you multiply by 2: 4 times 8 fourths is 32 fourths. That's 8.

f. So 4 times 8 fourths equals 8. Say that equation. (Signal.) *4 times 8 fourths equals 8.*

g. Remember how the fraction works:
• The number **after** the equal sign is always the **numerator** of the fraction.
• The **first number** in the problem is always the **denominator** of the fraction.

h. The same rules work for any multiplication problem.
• The equation in the second box has the fraction missing. You'll figure out that fraction: 5 times some fraction equals 2.
• The top arrow shows that the number after the equal sign is the numerator. What number? (Signal.) *2.*

• The other arrow shows that the first number in the problem is the denominator. What number? (Signal.) *5.*
• So 5 times what fraction equals 2? (Signal.) *2 fifths.*

i. Yes, 5 times **2 fifths** equals 2. Say the equation. (Signal.) *5 times 2 fifths equals 2.*
• It's true. 5 times 2 fifths equals 10 fifths. That's 2.

j. I'll say problems with the missing fraction. You'll tell me the missing fraction.
• Listen: 4 times some fraction equals 11. Remember, 11 is the numerator. Everybody, what's the fraction? (Signal.) *11 fourths.*

k. Listen: 17 times some fraction equals 2. Everybody, what's the fraction? (Signal.) *2/17ths.*
• Listen: 100 times some fraction equals 3. What's the fraction? (Signal.) *3/100ths.*
• Listen: 14 times some fraction equals 1. What's the fraction? (Signal.) *1/14th.*

• (Repeat step k until firm.)

In lesson 18, students work problems of the form

$$5\,(p) = 28,$$

using the analysis. They figure out the fraction for P and write the simple equation below.

Here's part of the introduction from lesson 18:

| $6\;(r) = 11$ $6\left(\frac{11}{6}\right) = 11$ $\boxed{r = \frac{11}{6}}$ | **a.** $4\,(r) = 7$ **b.** $15\,(d) = 9$ **c.** $8\,(v) = 1$ |
|---|---|

b. The problem in the box is like problems you've worked, but the missing fraction is shown with a letter. The problem is **6 times R equals 11.**
• The equation is written below with a fraction for R. The numbers for the fraction are 11 in the numerator and 6 in the denominator. So the simple equation for the letter is **R equals 11 sixths.**

- What's the simple equation for R? **(Signal.)** *R equals 11 sixths.*

c. Remember, for the missing fraction:
  - The number after the equal sign is the numerator of the fraction.
  - The other number is the denominator.
- You write the simple equation showing the fraction the letter equals.

d. I'll read problem A: 4 times R equals 7.
- Tell me the fraction R equals. **(Signal.)** *7 fourths.*
  Yes, R equals 7 fourths.

e. I'll read problem B: 15 times D equals 9.
- Tell me the fraction D equals. **(Signal.)** *9/15ths.*
  Yes, D equals 9/15ths.

f. I'll read problem C: 8 times V equals 1.
- Tell me the fraction V equals. **(Signal.)** *1 eighth.*
  Yes, V equals 1 eighth.

- (Repeat steps d–f until firm.)

> **Teaching note:** The series of tasks presented in steps d–f should not require a great deal of thinking time. Permit possibly two seconds for students to figure out the fraction. If they require much more time than that, they have not learned the procedure. Repeat d–f until they are able to say the correct fractions with no more than about a 2-second pause. Also, make sure the students respond together, on signal. Without this constraint, some students will simply copy the responses other students initiate.

Following the introduction, students copy each problem and write the simple equation below. Again, they should be able to work quickly.

In lessons 19 and 20, students work with problem sets that have problems with a missing factor and problems with a missing product.

$$4\left(\frac{3}{5}\right) = m, \quad 4\,(p) = 9$$

Students solve problems with a missing factor by multiplying, simplifying the answer, and writing a simple equation that shows what the letter equals.

In lessons 19–21, students generate multiplication equations from fractions.

Here's the introduction:

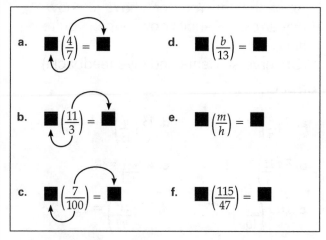

- These problems show the fraction you multiply by, but they don't show the rest of the equation.

b. What's the fraction for the first problem? **(Signal.)** *4/7.* Yes, 4/7.
- The arrows show where the numbers go in the equation. The **4** goes after the equal sign. **7** is the first number. The equation for that fraction is 7 × 4/7 = 4.
- Everybody, say the equation. **(Signal.)** *7 × 4/7 = 4.*

c. Read fraction B. **(Signal.)** *11/3.*
- Here's the equation for 11/3: 3 × 11/3 = 11.
- Say that equation. **(Signal.)** *3 × 11/3 = 11.*

d. Read fraction C. **(Signal.)** *7/100.*
- What will the equation start with? **(Signal.)** *100.*
- Say the equation. **(Signal.)** *100 × 7/100 = 7.*
  (Repeat step d until firm.)

e. The rest of the items do not have arrows. Remember, the top number goes after the equal sign. So the first number you say is the bottom number.

f. Read fraction D. **(Signal.)** *B over 13.*
- Say the equation. **(Signal.)** *13 × B/13 = B.*

g. Read fraction E. **(Signal.)** *M over H.*
- Say the equation. **(Signal.)** *H × M/H = M.*

h. Read fraction F. **(Signal.)** *115/47.*
- Say the equation. **(Signal.)** *47 × 115/47 = 115.*

- (Repeat steps f–h until firm.)

**i.** Your turn: Write the complete equation for each item. You don't have to show any arrows. Pencils down when you're finished.

(Observe students and give feedback.)

**Key:**

**a.** $7\left(\dfrac{4}{7}\right) = 4$      **d.** $13\left(\dfrac{b}{13}\right) = b$

**b.** $3\left(\dfrac{11}{3}\right) = 11$      **e.** $h\left(\dfrac{m}{h}\right) = m$

**c.** $100\left(\dfrac{7}{100}\right) = 7$      **f.** $47\left(\dfrac{115}{47}\right) = 115$

> **Teaching note:** If students have trouble with the verbal series of problems, steps f–h, repeat the entire series. Remind students that this exercise follows the same basic rules as the problems for which they identify the fraction.
>
> The number in the denominator is the first value in the equation. The number in the numerator is the number after the equal sign.

In lessons 22–24, students work with mixed sets of problems. Some give the fraction; some have a missing fraction.

Here's the problem set from lesson 23:

**a.** $115\left(\dfrac{\blacksquare}{\blacksquare}\right) = n$    **c.** $\blacksquare\left(\dfrac{1}{95}\right) = \blacksquare$    **e.** $\blacksquare\left(\dfrac{58}{3}\right) = \blacksquare$

**b.** $\blacksquare\left(\dfrac{b}{p}\right) = \blacksquare$    **d.** $jd\left(\dfrac{\blacksquare}{\blacksquare}\right) = x$

> **Teaching note:** The problems include both numbers and letters. The idea is for students to understand that the same procedure used for numbers applies to letters.

## ANALYSIS OF WORDING

Before students work entire word problems, they learn how to analyze the wording of the problem and translate it into an equation with letters. The first wording discrimination appears in lesson 17. Subsequent lessons continue to work on wording until lesson 26, when students solve complete problems.

In lessons 17 and 18, students work with questions that name two units. Students identify the unit name that answers the question. In these examples, the unit name follows the question-asking words How many . . . ? and What's the number of . . . ?

**a.** Listen: The word problems you'll work need a unit name in the answer.
- To answer the question "How many," you find the first unit name that follows the words **how many.**
- So if a question asks **how many books,** the unit name in the answer is **books.**
- If it asks **how many loads,** what's the unit name? (Signal.) *Loads.*
- If it asks **how many years,** what's the unit name? (Signal.) *Years.*

**b.** Here's a question: How many miles will the car go in 4 hours?
- What's the first unit name after **how many?** (Signal.) *Miles.*
  It's miles. So that's the unit name in your answer.

**c.** New question: How many days will it take to fill 56 barrels?
- What's the first unit name after **how many?** (Signal.) *Days.*
  Yes, days. That's the unit in the answer.

d. New question: If they eat 57 bushels, how many days will it take?
- What's the first unit name after **how many?** (Signal.) *Days.*
- So what's the unit name in the answer? (Signal.) *Days.*
  Yes, days.
e. New question: If they eat 58 pounds, how many hours will it take?
- What's the unit name in the answer? (Signal.) *Hours.*
f. New question: After she works 7 hours, how much money will she earn? The unit name is dollars.
- What's the unit name in the answer? (Signal.) *Dollars.*
  Yes, the question asks about money, but you know that the unit will be dollars.
g. New question: How many pounds would 12 barrels weigh?
- What's the unit name in the answer? (Signal.) *Pounds.*
h. New question: How many hours would 7 trips take?
- What's the unit name in the answer? (Signal.) *Hours.*
- (Repeat steps d–h until firm.)

Students next write unit names for eight new examples.

*Teaching note:* Although the discrimination seems very obvious, it is not obvious for the students. When you teach students, remember the procedure. When students later make mistakes, remind them of the rule:

What's the unit that comes right after **how many** in that question?

That's the unit the problem asks about.

In lesson 19, students learn the procedure for writing a complete letter equation from a question.

Here's part of the exercise from lesson 19. This part comes after the explanation of how to write the equation with letters. The explanation involves simple equations that have the same notation on both sides of the equal sign: $4 = 4$, $n = n$, and so forth.

---

a. How many houses are on 12 blocks?

b. If James exercises for 3 days, how many miles does he jog?

c. How many chairs are in 16 rows?

---

b. Question A. I'll write the equation on the board.
- The question asks: How many houses are on 12 blocks?
- What's the unit name in the answer? (Signal.) *Houses.*
- What's the **simple** equation I write? (Signal.) *H equals H.*
- (Write on the board:)                    [19:4A]

> **a.**        $h = h$

- Listen to the question again: How many houses are on 12 blocks?
- What's the letter for the other unit name? (Signal.) *B.*
- That letter goes in the other two places.
- (Write to show:)                         [19:4B]

> **a.**    $b \left( \dfrac{h}{b} \right) = h$

- Here's the complete equation: B × H over B = H.
- Say the equation for question A. (Signal.) *B × H over B = H.*
c. Your turn: Copy the complete equation for question A. √

**d.** Question B: If James exercises for 3 days, how many miles does he jog?
- What's the unit name in the answer? (Signal.) *Miles.*
- What's the other unit name? (Signal.) *Days.*
- The unit name in the answer is miles. Use the letter M for miles.
- Say the simple equation you start with. (Signal.) *M = M.*
- (Repeat step d until firm.)

**e.** Write the simple equation. Then complete the equation with the letter for the other unit. Pencils down when you've completed the equation.
(Observe students and give feedback.)

**f.** Check your work.
- (Write on the board:) [19:4C]

$$\textbf{b.} \qquad d\left(\frac{m}{d}\right) = m$$

- Here's the equation you should have for question B: If James exercises for 3 days, how many miles does he jog?
- What's the simple equation you start with? (Signal.) *M = M.*
- What's the other letter? (Signal.) *D.*
- Read the complete equation. (Signal.) *D (M/D) = M.*

**g.** Question C: How many chairs are in 16 rows?
- What's the unit name in the answer? (Signal.) *Chairs.*
- What's the other unit name? (Signal.) *Rows.*
- The unit name in the answer is chairs, so what's the simple equation you start with? (Signal.) *C = C.*
- Then you complete the equation. Pencils down when you've written the complete equation for question C.
(Observe students and give feedback.)
- Check your work.

- (Write on the board:) [19:4D]

$$\textbf{c.} \qquad r\left(\frac{c}{r}\right) = c$$

- Here's the equation you should have for the question: How many chairs are in 16 rows?
- What's the simple equation you start with? (Signal.) *C = C.*
- What's the other letter? (Signal.) *R.*
- Read the complete equation. (Signal.) *R times C over R equals C.*

This equation form,

$$b\left(\frac{h}{b}\right) = h, \qquad r\left(\frac{c}{r}\right) = c,$$

has been established in earlier lessons. Students now apply the same equation form to unit names.

**Teaching note:** Students must achieve a high level of mastery on this procedure. The goal is for them to become automatic with writing the equation. They will use this equation-construction procedure throughout the program. The practice that students receive in working only on the step of writing the equation prepares them for what they'll do later when they solve complete word problems. At that time, you should not have to spend time teaching them how to write the letter equations.

In subsequent lessons, students continue to practice writing equations with letters as they learn how to write equations with numbers for letter equations.

In lessons 22–24, students identify the numbers that go in the *fraction*. This discrimination is difficult for students. Generally, the two numbers for the fraction are in a single sentence that relates the unit: 5 miles every 3 days, 4 windows in every 2 rooms, 20 cups of sugar for every 30 cups of flour.

The items that students work show a complete word problem. The letter equation is given. Students identify the phrase that names the related units and write the corresponding letter equation with numbers for the fraction.

Here's part of the exercise from lesson 22:

| | | |
|---|---|---|
| a. | Every 4 hours, a factory produces 27 cars. How many cars does the factory produce in 11 hours? | $h\left(\dfrac{c}{h}\right) = c$ |
| b. | Robert grades 45 papers. He grades 15 papers every 4 minutes. How many minutes does it take to grade the 45 papers? | $p\left(\dfrac{m}{p}\right) = m$ |
| c. | 3 teams have a total of 18 players. If there are 72 players, how many teams are there? | $p\left(\dfrac{t}{p}\right) = t$ |
| d. | In 3 hours, Molly earns 42 dollars. How many hours would it take her to earn 100 dollars? | $d\left(\dfrac{h}{d}\right) = h$ |
| e. | In a room, there are 96 chairs. There are 12 chairs for every 2 tables. How many tables are in the room? | $c\left(\dfrac{t}{c}\right) = t$ |

- Each item shows a problem and the equation with letters. You're going to find the sentence that tells about related units and write a correct fraction for the problem.
- b. Item A: Every 4 hours, a factory produces 27 cars. How many cars does the factory produce in 11 hours?
- Raise your hand when you know the sentence that tells about the related units. √
- Everybody, read the sentence. (Signal.) *Every 4 hours, a factory produces 27 cars.*
- Copy the equation. Below, write the same equation with the numbers that show the relationship. Remember to put the related number for cars in the numerator and the related number for hours in the denominator. Pencils down when you're finished.
  (Observe students and give feedback.)

- (Write on the board:) [22:4A]

$$a. \quad h\left(\frac{c}{h}\right) = c$$

$$h\left(\frac{27}{4}\right) = c$$

- Here's what you should have.
- Everybody, read the **fraction** with numbers. (Signal.) *27/4.*
- c. Problem B: Robert grades 45 papers. He grades 15 papers every 4 minutes. How many minutes does it take to grade the 45 papers?
- Raise your hand when you know the sentence that tells about the related units. √
- Everybody, read the sentence. (Signal.) *He grades 15 papers every 4 minutes.*
- Copy the letter equation. Below, write the same equation with the numbers that show the relationship. Pencils down when you're finished.
  (Observe students and give feedback.)
- Check your work.
- Look at the number fraction. What's the numerator of the fraction? (Signal.) *4.*
- What's the denominator? (Signal.) *15.*

*Teaching note:* For some of these problems, the word problem names the related units in one order, but the numbers for the fraction are in the other order. For instance, problem A names the related units in the order hours and cars; however, hours is in the denominator.

Less structure is provided for the second problem than for the first. Make sure you follow the script closely because you don't want to provide any more structure than the students need. Following problem C, students work problems D and E without structure. If students make mistakes with the less structured problems, you can present the steps used for problem B.

The wording of the sentences is simple and lacks some of the qualifications later problems have. For instance, problem A refers to three teams but does not indicate that they are identical. These qualifications will occur in later problems.

In lesson 26, students work entire problems. They write the equation with letters; they write the complete equation with numbers (not simply the fraction); they solve the problem; and they write the answer to the question the problem asks. The answer is a number and a unit name. Below are the first three items from the problem set that appears in lesson 26.

> **a.** There are 10 chairs for every 4 tables. There are 140 chairs in all. How many tables are there?
>
> **b.** If there are 200 sinks in each hospital wing, how many sinks are needed for 12 wings?
>
> **c.** John runs 28 miles. He runs 14 miles every 4 days. How many days does he run?

Here's the first part of the presentation from lesson 26:

**b.** Problem A: There are 10 chairs for every 4 tables. There are 140 chairs in all. How many tables are there?
- What does the question ask about? (Signal.) *Tables.*
- Write the equation with letters. Then stop. (Observe students and give feedback.)
- (Write on the board:) [26:2A]

$$\textbf{a.} \qquad c\left(\frac{t}{c}\right) = t$$

- Here's what you should have.
- Below, you'll write an equation with 3 numbers.
- One of the sentences tells the relationship. Read that sentence. (Signal.) *There are 10 chairs for every 4 tables.*
- Write those numbers in the equation below. (Observe students and give feedback.)

- (Write to show:) [26:2B]

$$\textbf{a.} \qquad c\left(\frac{t}{c}\right) = t$$
$$\left(\frac{4}{10}\right) = t$$

- Here's what you should have so far.
- One of the sentences gives the number for either chairs or tables. Read that sentence. (Signal.) *There are 140 chairs in all.*
- Write that number in the equation. √
- (Write to show:) [26:2C]

$$\textbf{a.} \qquad c\left(\frac{t}{c}\right) = t$$
$$140\left(\frac{4}{10}\right) = t$$

- Here's what you should have.
- Skip 3 lines so you can finish the problem later. √

*Teaching note:* During the structured part shown above, students do not work the whole problem. After they write equations for B, C, and D, they go back and work the items one at a time. This format provides students with fewer interruptions on the new part of the strategy they are learning. Instead of working a whole problem before going to the next example, they work only the critical part, writing the letter equation and corresponding number equation.

In lessons 27 through 30, the students continue to work problems of this type, but with less structure. Also, they work a complete problem before working the next problem. At this time, students should be facile with the steps in working a problem; however, there are many steps, and you may have to remind them about writing the answer to the question the problem asks and showing the unit name in the answer.

Do not permit students to take shortcuts. It is better for them to follow all the steps quickly and take a little longer to complete the problem than it is to take shortcuts that save a little time now but will be very costly later when students are required to solve more difficult problems for which shortcuts will likely lead to errors.

In lessons 31–36, the program introduces variations in wording. One variation presents a question that does not refer to the unit name:

> **How long will it take Joe to lose 25 pounds if he loses 2 pounds every 3 weeks?**

Another variation presents problems that refer to subsets, for instance, red balls and yellow balls.

> **There are 15 red balls for every 46 yellow balls. If there are 60 red balls, how many yellow balls are there?**

Students write equations with a two-letter designation for each subtype: RB and YB.

## NOTE CONCERNING MATHEMATICAL NOTATION

Some teachers who field-tested *Essentials* had two concerns about using double-letter notations like *rb* for red balls and *yb* for yellow balls. The concerns were that students would interpret *rb* as *r* times *b* and that using *rb* to represent a single variable was mathematically incorrect because *rb* means *r* times *b* in algebraic equations.

Field-test results indicate that students will not confuse double letters representing one variable in a rate equation with two letters representing two separate variables that are multiplied.

We can also infer that students who are firm on constructing and solving rate-equation problems won't confuse the double-letter notation with two letters multiplied together.

Students generate the equations containing the double-letter notation from a word problem, so they understand what the double letters represent. Students won't confuse *rb* with *r* times *b* because only one number is given for *rb*, not one number for *r* and another for *b*.

The double-letter notation is mathematically appropriate as well. For example, in calculus equations written in the Leibniz notation, letter pairs starting with *d* and followed by another letter represent the derivative of the other letter *(dx, dy, dz, du)*. The letter pairs describe a single function and not the individual letters multiplied together.

In lesson 37, the program introduces "ratio" language, such as *The ratio of boys to girls is 5 to 7*. Students learn the correspondence between the order of names and the order of numbers. Students first go over a problem that is written and explained. Then they identify the numbers that correspond to the words.

Here's that part of the exercise from lesson 37:

---

◆ The ratio of dogs to cats is 5 to 9.

◆ The ratio of men to women is 4 to 3.

◆ Sand and gravel are mixed in the ratio of 5 to 4.

◆ In a garden, the ratio of blue flowers to red flowers is 8 to 3.

◆ The ratio of broken shells to shells that are not broken is 15 to 4.

---

c. Here's a ratio: The ratio of dogs to cats is 5 to 9.
- Say that sentence. (Signal.) *The ratio of dogs to cats is 5 to 9.*
- The first number is 5. What animal does that refer to? (Signal.) *Dogs.*
  And the 9 refers to cats.

d. New ratio: The ratio of men to women is 4 to 3.
- Say that sentence. (Signal.) *The ratio of men to women is 4 to 3.*
- The first number is 4. What does that refer to? (Signal.) *Men.*
- And what does the 3 refer to? (Signal.) *Women.*

e. Sand and gravel are mixed in the ratio of 5 to 4.
- Which number tells about sand? (Signal.) *5.*
- Which number tells about gravel? (Signal.) *4.*

f. In a garden, the ratio of blue flowers to red flowers is 8 to 3.
- Which number tells about blue flowers? (Signal.) *8.*
- Which number tells about red flowers? (Signal.) *3.*

g. The ratio of broken shells to shells that are not broken is 15 to 4.
- Which number tells about shells that are not broken? (Signal.) *4.*
- (Repeat steps e–g until firm.)

In lesson 41, students discriminate between problems that call for a rate equation and those that call for an addition equation that involves two subsets, a type taught in lessons 22 through 27. (See page 90.)

Below is the first part of the exercise from lesson 41:

---

**a.** Men and women volunteer at the library in the ratio of 2 to 5. There are 40 women on the volunteer list. How many men are on the list?

**b.** There are 64 adults in the city library. There are 40 women in the library. How many men are in the library?

**c.** There are lots of vehicles in the warehouse. 56 of the vehicles are new. 85 of the vehicles are used. How many vehicles are in the warehouse?

**d.** There are 5 used vehicles for every 3 new vehicles on the sales lot. If there are 90 used vehicles, how many new vehicles are on the lot?

**e.** Students choose either hot or cold lunch each day. On Wednesday, a total of 285 lunches were chosen. 212 students chose hot lunch. How many students chose cold lunch on Wednesday?

---

- You'll work some of these problems using a rate equation. For others, you'll add or subtract. For all of them, you'll write the answer with a number and a unit name.
b. I'll read each problem. Tell me if it's **rate** or **add-subtract.**

c. Problem A: Men and women volunteer at the library in the ratio of 2 to 5. There are 40 women on the volunteer list. How many men are on the list?
- Raise your hand when you know if this is a rate or an add-subtract problem. √
- Everybody, is it a rate problem or an add-subtract problem? (Signal.) *Rate problem.*

d. Problem B: There are 64 adults in the city library. There are 40 women in the library. How many men are in the library?
- Raise your hand when you know if this is a rate or an add-subtract problem. √
- Everybody, is it a rate problem or an add-subtract problem? (Signal.) *Add-subtract problem.*

e. Problem C: There are lots of vehicles in the warehouse. 56 of the vehicles are new. 85 of the vehicles are used. How many vehicles are in the warehouse?
- Raise your hand when you know if this is a rate or an add-subtract problem. √
- Everybody, is it a rate problem or an add-subtract problem? (Signal.) *Add-subtract problem.*

After students respond verbally to the rest of the problems in the set, they write the letter equation for each problem. After writing all the equations, students work all the problems.

---

*Teaching note:* Two pairs of problems are minimally different. Within each pair, the same names appear (for example, new and used vehicles). If students are firm on what they have learned through lesson 40, they should not have difficulties with this set. Note, however, that some students do not attend to all the details of the problem. Often they will try to set up a rate equation with letters based on the question the problem asks. Usually, they discover that there are not related units (so they can't write the fraction) and realize that the problem can't be a rate problem.

Remind them that if they find a sentence that tells about the related units, they write a rate equation. If the problem doesn't have a sentence that shows related units, it is an addition-subtraction problem.

---

In lessons 45–47, students work on other kinds of mixed sets. They also work problems that refer to **per** and **an** (6 bottles per carton, 5 miles an hour).

Rate equations written in reverse order are introduced in lessons 49–52. Students still construct the equation based on the question the problem asks, but instead of writing the equation

$$h\left(\frac{m}{h}\right) = m,$$

they write

$$m = \left(\frac{m}{h}\right) h.$$

The main reasons for introducing the reverse order are

1. Students will use this order when they work with straight-line equations on the coordinate system:

$$y = \left(\frac{y}{x}\right) x.$$

2. This order parallels the way they write multiplication equations in the Algebra Translation track. (See page 89.)

Students start with the unit named in the question and write a simple equation. Then they complete the equation with the letter for the other unit written on the right side of the equal sign.

Here is the procedure from lesson 49 that students follow for the word problems.

---

◆ The ratio of peanuts to all nuts is 3 to 7. If there are 66 peanuts, how many nuts are there in all?

    ◆ Start with **n** and write
      the simple equation:     $n = n$

◆ Complete the rate equation:     $n = \left(\frac{n}{p}\right) p$

---

- Which name does the problem ask about, **all nuts** or **peanuts?** (Signal.) *All nuts.*
- So we start with N and write the simple equation N = N.
- Next, we complete the rate equation.
- You can see that equation below the simple equation: N = N over P times P.
- Everybody, say the whole equation. (Signal.) *N = (N/P) P.*

i. Remember, start with the unit that answers the question. Make a simple equation, and then complete the rate equation.
- For each problem, you'll write the equation so that the unit that answers the question comes first, followed by an equal sign.

b. Problem A: There are 6 dogs for every 15 cats. If there are 135 cats, how many dogs are there?
- What unit does the problem ask about? (Signal.) *Dogs.*
- (Write on the board:)     [49:1A]

> **a.**     $d = d$

- Write the simple equation **D = D.** Then complete the part of the rate equation that comes **after the equal sign.** Pencils down when you've written the letter equation.
(Observe students and give feedback.)
- (Write to show:)     [49:1B]

> **a.**     $d = \left(\frac{d}{c}\right) c$

- Here's the complete equation you should have: D = D over C times C.

*Teaching note:* Students may wonder why you're requiring them to write the equations in a new order. The simplest answer is so they will be ready to handle the equation for straight lines on the coordinate system.

In lessons 53 and 54, students work a mixed set of algebra multiplication problems and rate problems. Students write rate problems in the new order.

In lessons 61–63, students solve problems that ask about rate. For these problems, students write a simple equation,

$$\frac{m}{t} = \frac{m}{t}.$$

First, students identify the fraction that is named by the question. Here's part of the exercise that follows the introduction in lesson 61:

| | |
|---|---|
| **a.** | How many feet per second did the bird fly? |
| **b.** | How many gallons per house did the painters need? |

e. Question A: How many feet per second did the bird fly?
- What unit answers the question? (Signal.) *Feet per second.*
- What's the letter fraction for that unit? (Signal.) *F over S.*
- So you would write **F over S** = for the first part of the equation.
- (Write on the board:)                    [61:2A]

$$\textbf{a.} \quad \frac{f}{s} =$$

- Here's what you'd write.
- Then you complete the simple equation: F over S = F over S.
- (Write to show:)                    [61:2B]

$$\textbf{a.} \quad \frac{f}{s} = \frac{f}{s}$$

- Write this equation for A. √
  f. Question B: How many gallons per house did the painters need?
- What unit answers the question? (Signal.) *Gallons per house.*
- What's the letter fraction for that unit? (Signal.) *G over H.*
- Say the simple equation. (Signal.) *G/H = G/H.*

In the same exercise, students write simple equations based on the questions:

i. Go back to question B: How many gallons per house did the painters need?
- Start with the fraction that answers the question and write the simple equation. Pencils down when you're finished. (Observe students and give feedback.)

- (Write on the board:)                    [61:2C]

$$\textbf{b.} \quad \frac{g}{h} = \frac{g}{h}$$

- Here's what you should have. G/H = G/H.
  j. Question C: How many houses per block did the company build?
- Start with the fraction that answers the question and write the simple equation. Pencils down when you're finished. √
- (Write on the board:)                    [61:2D]

$$\textbf{c.} \quad \frac{h}{b} = \frac{h}{b}$$

- Here's what you should have.

In lessons 62 and 63, students solve problems for the rate. Here's the first example they work:

*A snail was observed for 3 hours. The snail traveled 57 inches in that time. How many inches per hour did the snail travel?*

Students solve the problem for the fraction on the left side of the equation. Here are the steps:

- First, they write the equation with letters:

$$\frac{i}{h} = \frac{i}{h}$$

- Next, they put in the numbers the problem gives:

$$\frac{i}{h} = \frac{57}{3}$$

- Next, they solve for $\frac{i}{h}$ :

$$\frac{i}{h} = 19.$$

- Finally, they write the answer as a number and unit name: **19 inches per hour.**

In lessons 69–72, students work problems that have variations in the question. One variation names one, but not both, of the units: "What is his rate of filing papers? What is the train's rate per hour?"

The final variation does not name either unit: "What's the train's average rate?"

Lesson 72 presents a mixed set of problems: those that are solved with a rate equation and those that ask about the rate with different question forms.

## INTERPRETING TABLES AND GRAPHS (Lessons 94–101)

In lessons 94–101, students work from tables and graphs showing survey data. Students work problems that require them to extrapolate and predict outcomes. For example, students use information about preferred activities for a small population of students and make predictions about expected numbers for a larger population of students. Based on the question the problem asks, students refer to the table or graph, write a rate equation, and solve it.

The survey data provide ratio numbers that students put in the fraction of a rate equation.

Students first refer to the problem to identify the two units that will be in the equation with letters. For example, the problem asks, "If 560 students go to school on Tuesday, how many would you expect to ride a bicycle?" The question asks about riding a bicycle (b). The other unit is total students. So an acceptable equation would be

$$b = \left(\frac{b}{a}\right) a.$$

If the problem asks, "On Tuesday, how many students would either walk or ride in a car?" students write

$$w + c = \left(\frac{w + c}{a}\right) a.$$

Students refer to the table or graph to find the numbers that go in the fraction.

Here's the first part of the exercise from lesson 97:

| 2-hour observation | | | | | | | | | | | | | | | | | | | | | | | | | |
|---|---|---|---|---|---|---|---|---|---|---|---|---|---|---|---|---|---|---|---|---|---|---|---|---|---|
| tigers: | t | | | | | | | | | | | | | | | | | | | | | | | | |
| lions: | l | | | | | | | | | | | | | | | | | | | | | | | | |
| baboons: | b | | | | | | | | | | | | | | | | | | | | | | | | |
| crocodiles: | c | | | | | | | | | | | | | | | | | | | | | | | | |
| Times seen | | 1 | 2 | 3 | 4 | 5 | 6 | 7 | 8 | 9 | 10 | 11 | 12 | 13 | 14 | 15 | 16 | 17 | 18 | 19 | 20 | 21 | 22 | 23 | 24 | 25 |

**a.** If the animals were observed for 3 hours, how many lions would you expect to observe?

**b.** If you observe 48 tigers, about how many crocodiles would you expect to see in the same time period?

• This graph shows the number of tigers, lions, baboons, and crocodiles that were observed in a wildlife refuge. The numbers below show the number of animals observed in a 2-hour period.
• How many crocodiles were observed? (Signal.) *11.*
• How many lions were observed? (Signal.) *12.*
b. Item A: If the animals were observed for 3 hours, how many lions would you expect to observe?
• Work the problem. Remember, you get the numbers for the fraction from the information in the table. Pencils down when you're finished.
(Observe students and give feedback.)
• Everybody, read the equation with letters. (Signal.) *L = L over H times H.*
Yes, this problem asks about the number of lions. The other letter is for hours.
• The equation with numbers is L = 12/2 times 3.
• The numbers in the fraction are information the table gives. There are 12 lions in 2 hours.
• What's the answer to the problem? (Signal.) *18 lions.*
Yes, you'd expect to see 18 lions in 3 hours.
c. Item B: If you observe 48 tigers, about how many crocodiles would you expect to see in the same time period?
• Work the problem. Pencils down when you're finished.
(Observe students and give feedback.)
• Everybody, read the equation with letters. (Signal.) *C = C over T times T.*
• The equation with numbers is C = 11/16 times 48.
• What's the answer? (Signal.) *33 crocodiles.*

For some of the problems students work, they must combine values presented in tables. For instance,

*If you observed for 3 hours, how many baboons and crocodiles would you expect to observe?*

The equation is

$$b + c = \left(\frac{b+c}{h}\right) h.$$

To obtain a number for $b + c$, students combine numbers shown in the table: $21 + 11$.

Later problems include circle graphs with percents and matrices, such as responses to different questions on an opinion poll.

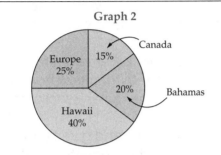

**Graph 2**

◆ A survey was taken to ask which of 4 travel destinations people preferred. Percentages for the 4 choices are displayed in Graph 2 above.

   a. If 150 people took part in the survey, how many of them selected either Europe or Canada as a travel destination?

   b. If 750 people were asked to select from 4 travel destinations, how many of them would you expect to choose the Bahamas?

   c. If 80 people select Europe, about how many would select Hawaii?

   d. If 128 people choose to go to Hawaii, about how many people in all were asked to choose a destination?

   e. If 260 people were asked to choose a destination, about how many people would you expect to choose Canada?

## UNIT CONVERSION (Lessons 113–116)

Starting in lesson 113, students work unit-conversion problems. The first type of problem that students work asks about the conversion.

> **How many inches are in 13 feet?**

The basic procedure that students follow is to write a rate equation for related units.

Here's part of the introduction from lesson 113:

| | |
|---|---|
| 1 foot = 12 inches<br>1 yard = 3 feet = 36 inches<br>1 mile = 1760 yards = 5280 feet | 1 centimeter = 10 millimeters<br>1 meter = 100 centimeters<br>1 kilometer = 1000 meters |
| 1 pint = 2 cups = 16 fluid ounces<br>1 quart = 2 pints = 4 cups<br>1 gallon = 4 quarts = 8 pints | 1 pound = 16 ounces<br>1 ton = 2000 pounds<br>1 kilogram = 1000 grams |
| 1 week = 7 days<br>1 year = 12 months<br>1 year = 365 days | 1 minute = 60 seconds<br>1 hour = 60 minutes<br>1 day = 24 hours |

• To work the problems in part 4, you write a rate equation based on the question. The information you'll need for the related units is shown in the table.

b. Problem A: How many inches are in 13 feet?

• What unit does the problem ask about? (Signal.) *Inches.*

• What's the related unit? (Signal.) *Feet.*

• Write the equation with letters. Then stop. √

• (Write on the board:)      [113:4A]

    **a.**    $i = \left(\dfrac{i}{f}\right) f$

• Here's what you should have: I = I over F times F.

• The table shows how many inches are in a foot. Use those numbers in the fraction, and write the equation with numbers. Then stop. √

• (Write to show:)      [113:4B]

    **a.**    $i = \left(\dfrac{i}{f}\right) f$

         $i = \left(\dfrac{12}{1}\right) 13$

• Here's what you should have: I = 12/1 times 13.

• Skip 3 lines.

c. Problem B: How many feet are in 13 inches?

• Write the equation with letters. Below, write the equation with 3 numbers. Pencils down when you've done that much. √

- (Write on the board:) [113:4C]

$$f = \left(\frac{f}{i}\right)i$$

**b.**

$$f = \left(\frac{1}{12}\right)13$$

- Here's what you should have: F = F over I times I. The equation with numbers is F = 1/12 times 13.
- Skip 3 lines.

*Teaching note:* By keying the equation to the question, students do not have to follow mechanical rules about multiplying the larger unit to obtain the smaller unit or dividing the smaller unit to obtain the larger. If the question asks how many inches are in 11 feet, students set up the equation that shows what I equals. If the question asks how many feet are in 78 inches, students set up the equation that shows what F equals.

In lesson 115, students work word problems that name two convertible units and therefore require unit conversion. For example:

> *It takes a crew 2 weeks to lay 1.5 miles of pavement. How much pavement can the crew lay in 20 days?*

The problem gives two units for time—weeks and days. In this context, students always convert the larger unit into the smaller. The smaller unit is days, so students write the equation

$$d = \left(\frac{d}{w}\right)w.$$

After converting weeks into days, students write the rate equation for solving the problem,

$$m = \left(\frac{m}{d}\right)d,$$

and figure out the answer.

*Teaching note:* If students become confused about working problems with two convertible units, remind them that they are to end up with the smaller unit. Tell them to touch the smaller unit in the problem. Remind them that they want to end up with that unit. So they write the equation that tells what that unit equals.

## Fraction Simplification (Lessons 16–29)

Starting in lesson 16, this track introduces divisibility rules for 10, 5, 2, 3, 6, and 9 and procedures for identifying values divisible by multiples of 2. Students simplify fractions by applying the divisibility rules. Students will continue to simplify fractions throughout the program.

Here are the divisibility rules and the lessons in which they are introduced.

Lesson 16

> If the number is divisible by 10, the last digit is zero.

> If the number is divisible by 5, the last digit is 5 or 0.

> If the number is divisible by 2, the last digit is even.

Lesson 19

> If the number is divisible by 3, the sum of the digits is a multiple of 3.

Lesson 24

> If the number is divisible by 3 and 2, it is divisible by 6.

Lesson 27

> If the number is divisible by 9, the sum of the digits is a multiple of 9.

The general presentation format is that following the introduction of each rule, students apply the rule first to whole numbers, then to fractions.

In lesson 17, students learn the basic rule about simplifying fractions: **Divide both numerator and denominator by the same value.** Students also learn to divide by the largest possible value first.

In lesson 18, students practice with fractions that can be simplified more than once.

Here's the exercise:

◆ If you can divide by more than one value, first divide by the largest value.  $\frac{20}{60} = \frac{2}{6} = \boxed{\frac{1}{3}}$

◆ You'll find some fractions that can be simplified again.

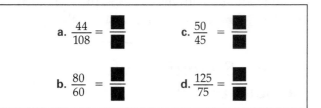

a. $\frac{44}{108} = \frac{\blacksquare}{\blacksquare}$    c. $\frac{50}{45} = \frac{\blacksquare}{\blacksquare}$

b. $\frac{80}{60} = \frac{\blacksquare}{\blacksquare}$    d. $\frac{125}{75} = \frac{\blacksquare}{\blacksquare}$

b. You've learned about simplifying fractions. You divide both the numerator and the denominator by the same value. Remember the rule in the box:
  • If you can divide by more than one value, first divide by the largest value.
  • When you do this, you'll find that some fractions can be simplified again.
c. You can see the fraction: 20/60ths.
  • Is the largest value you can divide both numbers by 10, 5, or 2? (Signal.) *10.*
  • When you divide by 10, you get 2 sixths.
  • Then you look at 2 sixths to see if **that** fraction can be simplified. Can 2 sixths be simplified? (Signal.) *Yes.*
  • What can you divide 2 and 6 by? (Signal.) *2.*
  • What fraction do you get? (Signal.) *1 third.* So the equation is 20/60ths equals 2 sixths equals 1 third.
d. Remember, if a fraction can be simplified more than once, write another equal sign and the simplified fraction.
e. Look at fraction A: 44 over 108.
  • Are both numbers divisible by 10? (Signal.) *No.*
  • By 5? (Signal.) *No.*
  • By 2? (Signal.) *Yes.*
  • Copy the fraction. Write an equal sign and the fraction you get when you divide 44 and 108 by 2. Pencils down when you've done that much. √
  • Everybody, what fraction did you get? (Signal.) *22 over 54.*
  • Can you simplify that fraction? (Signal.) *Yes.*

• Write another equal sign and the simplified fraction. Pencils down when you're finished. √
• Everybody, what's the simplified fraction? (Signal.) *11 over 27.*
• (Write on the board:)                    [18:5A]

a.    $\frac{44}{108} = \frac{22}{54} = \boxed{\frac{11}{27}}$

• Check your work. Here's what you should have.
f. Simplify the rest of the fractions in part 6.
• Remember, if the fraction is divisible by 10 and 2, first divide by 10. Then divide by 2. Pencils down when you're finished. (Observe students and give feedback.)
g. Check your work.
• (Assign 3 students to show the work for problems B–D on the board. Students are to work simultaneously.)

***Key for board:***

b. $\frac{80}{60} = \frac{8}{6} = \boxed{\frac{4}{3}}$    c. $\frac{50}{45} = \boxed{\frac{10}{9}}$

d. $\frac{125}{75} = \frac{25}{15} = \boxed{\frac{5}{3}}$

• Here's what you should have.
h. Fraction B: 80/60ths. First you divided both numbers by 10.
• What fraction did you get? (Signal.) *8 sixths.*
• What is 8 sixths divisible by? (Signal.) *2.*
• So what's the simplified fraction? (Signal.) *4 thirds.*
i. Fraction C: 50 over 45. You divided both numbers by 5.
• What's the simplified fraction? (Signal.) *10 ninths.*
j. Fraction D: 125 over 75. First you divided both numbers by 5.
• What fraction did you get? (Signal.) *25/15ths.*
• What is 25/15ths divisible by? (Signal.) *5.*
• So what's the simplified fraction? (Signal.) *5 thirds.*

By lesson 29, the last day of the track, students work with a full mix of fractions with a minimum of structure or leading from the teacher. The set consists of fractions that can be simplified once or more than once by the various numbers: 10, 2, 5, 3, 6, 9.

| | | |
|---|---|---|
| a. $\frac{45}{99}$ | c. $\frac{52}{48}$ | e. $\frac{90}{120}$ |
| b. $\frac{300}{250}$ | d. $\frac{81}{126}$ | f. $\frac{425}{950}$ |

## Fraction Simplification in Lessons 24–30

These lessons teach students the procedures for simplifying fractions in multiplication problems, then simplification for fractions that are added or subtracted.

In lesson 24, students learn to simplify fractions in problems that have a whole number multiplied by a fraction. The simplified answer is a whole number.

Students learn the basic rule that **if values are multiplied, you can simplify across values.**

The procedure they follow early in the track is to simplify fractions first if possible, then simplify across values. Student work that does not follow this order should not be marked wrong so long as the final answer can't be simplified. Point out, however, that if students simplify across values, they won't have to multiply large values.

Here's the Answer Key for the exercise from lesson 24:

Note that students show the simplified values by crossing out the value to be simplified, then writing new numbers above or below the original. Students are expected to follow this procedure throughout the program.

In lesson 25, the set of problems is expanded to include a fraction times a fraction and answers that are not whole numbers. The set for Lesson 26 includes letters. Students learn that a letter over the same letter equals one and can be crossed out. A letter over a different letter cannot be crossed out.

Here's the Answer Key for lesson 26.

In lessons 27–30, students work with a mix of problems that add or subtract and problems that multiply. Students learn the rule that if you add or subtract, you cannot simplify across fractions.

Students apply the rules for adding or subtracting and for multiplying. Here's the exercise from lesson 29.

a. $\dfrac{3}{2} - \dfrac{6}{12} = $ ■       d. $6\left(\dfrac{9}{27}\right) = $ ■

b. $\dfrac{6}{18} + \dfrac{12}{9} = $ ■      e. $\dfrac{2a}{9}\left(\dfrac{3}{a}\right) = $ ■

c. $\dfrac{a}{6}\left(\dfrac{12}{10}\right) = $ ■      f. $\dfrac{2}{16} + \dfrac{3}{8} = $ ■

- Some of these problems multiply and some add or subtract. You can simplify across values if you multiply. You **can't** simplify across values if you add or subtract.
b. Problem A: 3/2 − 6/12.
- Are you multiplying? (Signal.) *No.*
- So can you simplify across fractions? (Signal.) *No.*
c. Problem B: 6/18 + 12/9.
- Are you multiplying? (Signal.) *No.*
- So can you simplify across fractions? (Signal.) *No.*
d. Problem C: A over 6 times 12/10.
- Are you multiplying? (Signal.) *Yes.*
- So can you simplify across fractions? (Signal.) *Yes.*
- (Repeat steps b–d until firm.)
e. Copy each problem and work it. Do any simplification and then add, subtract, or multiply. Pencils down when you're finished.
(Observe students and give feedback.)

**Key:**

a. $\dfrac{3}{2} - \dfrac{\cancel{6}^1}{\cancel{12}_2} = \dfrac{2}{2} = \boxed{1}$    d. $^2\cancel{6}\left(\dfrac{\cancel{9}^1}{\cancel{27}_3}\right) = \boxed{2}$

b. $\dfrac{\cancel{6}}{_3\cancel{18}} + \dfrac{\cancel{12}^4}{\cancel{9}_3} = \boxed{\dfrac{5}{3}}$    e. $\dfrac{2\cancel{a}}{\cancel{9}_3}\left(\dfrac{\cancel{3}^1}{\cancel{a}}\right) = \boxed{\dfrac{2}{3}}$

c. $\dfrac{a}{\cancel{6}}\left(\dfrac{\cancel{12}^{\cancel{6}}}{\cancel{10}_5}\right) = \boxed{\dfrac{a}{5}}$    f. $\dfrac{\cancel{2}}{_8\cancel{16}} + \dfrac{3}{8} = \dfrac{4}{8} = \boxed{\dfrac{1}{2}}$

f. Check your work. Find part K at the end of lesson 29 in your textbook. That shows what you should have for problems A through F.
(Observe students and give feedback.)

*Teaching note:* In step a, students first review the rule about simplifying across values. In steps b–d, they apply the rule to verbal examples.

Finally, students work all the problems without additional structure. By lesson 30, students should be reliable in applying correct procedures for simplifying different types of problems.

# Algebra (Lessons 19–59)

The basic Algebra track teaches students how to solve one-step problems, two-step problems, problems that require substitution, and problems that require combining like terms. Students learn to solve for a letter that is on either the left or right side and learn to simplify expressions that have letter or number terms.

## ONE-STEP PROBLEMS (19–29)

The first exercises (lessons 19–21) present the letter term on the left. Students add or subtract on both sides to get the letter alone on the side.

In lesson 20, students learn that if they change the value on one side of the equation, they must change the other side in the same way.

Students first identify the letter they want to get alone. Then they indicate how they change both sides.

Here's part of the introduction from lesson 20:

b. Here's a rule about changing a side of an equation: If you change **one** side, you must change the **other** side in the **same way.**
c. Everybody, say that rule. (Signal.) *If you change one side, you must change the other side in the same way.*
- (Repeat step c until firm.)
d. If you change **one side** by subtracting 3, you must subtract 3 from the **other side.**
- If you change one side by adding 5, you must add 5 to the other side.
e. If you subtract 3 from one side, how do you change the other side? (Signal.) *Subtract 3.*
- If you add 5 to one side, how do you change the other side? (Signal.) *Add 5.*
- (Repeat step e until firm.)

Students next apply the procedure to workbook problems. They first identify the unknown, then how they need to change both sides to get the unknown alone.

| | | |
|---|---|---|
| a. $k - 2 = 23$ | | d. $t - 30 = 1$ |
| b. $17 + r = 24$ | | e. $\frac{3}{5} + y = \frac{5}{5}$ |
| c. $g + 13 = 17$ | | |

m. Look at problem A. √
  • Read problem A. (Signal.) *K − 2 = 23.*
  • What do you want to get alone? (Signal.) *K.*
  • How do you change both sides? (Signal.) *Add 2.*
n. Read Problem B. (Signal.) *17 + R = 24.*
  • What do you want to get alone? (Signal.) *R.*
  • How do you change both sides? (Signal.) *Subtract 17.*
o. Read Problem C. (Signal.) *G + 13 = 17.*
  • What do you want to get alone? (Signal.) *G.*
  • How do you change both sides? (Signal.) *Subtract 13.*
p. Read problem D. (Signal.) *T − 30 = 1.*
  • What do you want to get alone? (Signal.) *T.*
  • How do you change both sides? (Signal.) *Add 30.*
q. Read problem E. (Signal.) *3/5 + Y = 5/5.*
  • What do you want to get alone? (Signal.) *Y.*
  • How do you change both sides? (Signal.) *Subtract 3/5.*
  • (Repeat steps m–q until firm.)

After students tell the steps they will take, they work and check problem A, then work the rest of the problems.

**Teaching note:** If students are firm on the verbal tasks, they will make fewer mistakes when they actually perform the operation of changing both sides and writing the resulting equation.

Repeat any step in which students give weak responses. If students make mistakes, repeat the entire series, starting with problem A.

Observe students as they work the written problems that follow. Do not permit shortcuts. Students are to show the original equation and the value added or subtracted on both sides. Then they write a simple equation for the letter and box the equation.

Lessons 22 and 23 introduce problems that have the letter on either the left or the right side. Here's the problem set from lesson 23:

| | | |
|---|---|---|
| a. $\frac{7}{3} = \frac{2}{3} + j$ | | d. $\frac{2}{5} = m - \frac{9}{5}$ |
| b. $r - 51 = 100$ | | e. $r + 65 = 70$ |
| c. $13.6 + b = 25.8$ | | f. $6.05 = j + 5.9$ |

The set of examples varies with respect to which side the letter appears on, whether a value is added or subtracted, and whether the calculation involves whole numbers, fractions, or decimals.

In lesson 24, students learn to multiply by the reciprocal to get a letter alone on a side.

Students first learn that the reciprocal is a value turned upside down and that multiplying any value by its reciprocal equals 1.

Here's the introduction from lesson 24.

◆ Any value turned upside down is a value's reciprocal.

◆ The reciprocal of $\frac{2}{3}$ is $\frac{3}{2}$.

b. You're going to learn about reciprocals. Say reciprocal. (Signal.) *Reciprocal.*
  • Listen: Any value turned upside down is the value's reciprocal. The box shows an example. The reciprocal of 2/3 is 3/2.
c. What's the reciprocal of 2/3? (Signal.) *3/2.*
  • What's the reciprocal of 7/10? (Signal.) *10/7.*
  • What's the reciprocal of 1/9? (Signal.) *9 over 1.*
  That's 9.
  • (Repeat step c until firm.)

- (Teacher reference:)

> ◆ Any value multiplied by its reciprocal = 1.
>
> $\frac{5}{3} \times \frac{3}{5} = \frac{15}{15} = 1$    $\frac{5}{3} \times \frac{3}{5} = \frac{3}{3} \times \frac{5}{5}$
>
>                                     $= 1 \times 1 = 1$

d. The next box shows a rule you'll use to work a lot of problems with letters: Any value multiplied by its reciprocal equals 1.

- Listen again: Any value multiplied by its reciprocal equals 1.
- e. Say the rule with me. (Signal.) *Any value multiplied by its reciprocal equals 1.*
- Your turn: Say the rule. (Signal.) *Any value multiplied by its reciprocal equals 1.*
- (Repeat step e until firm.)

The exercise next provides the rationale for why a value times its reciprocal equals one. Then students apply the rule to a series of problems.

> a.    $\frac{3}{5} \times m = 6$      b.    $11 = g \times \frac{7}{2}$

e. Read problem A. (Signal.) *3/5 × M = 6.*
- Read everything on the side with the letter. (Signal.) *3/5 × M.*
- You're multiplying M by 3/5, so you multiply that side by the reciprocal of 3/5. What's the reciprocal of 3/5? (Signal.) *5/3.*
- Multiply both sides by 5/3. Below, write what M equals. Pencils down when you're finished.
(Observe students and give feedback.)
- (Write on the board:)                                    [24:1A]

> a.    $\left(\frac{5}{3}\right)\frac{3}{5} \times m = 6\left(\frac{5}{3}\right)$
>
>         $m = \left(\frac{30}{3}\right)$
>
>         $\boxed{m = 10}$

- Here's what you should have.
- On the side with the letter, you get 15/15 × M. That's 1 × M.

- On the other side, you get 30/3. So the simple equation is M = 10.
f. Read problem B. (Signal.) *11 = G × 7/2.*
- Read everything on the side with the letter. (Signal.) *G × 7/2.*
- You're multiplying G by 7/2.
- What's the reciprocal of 7/2? (Signal.) *2/7.*
- Multiply both sides by 2/7. Below, write the fraction that equals G. Pencils down when you're finished.
(Observe students and give feedback.)
- (Write on the board:)                                    [24:1B]

> b.    $\left(\frac{2}{7}\right)11 = g \times \frac{7}{2}\left(\frac{2}{7}\right)$
>
>         $\boxed{\frac{22}{7} = g}$

- Here's what you should have.
- On the side with the letter, you get G × 14/14.
- On the other side, you get 22/7. So the simple equation is 22/7 = G.

In lessons 26–28, students practice decomposing fractions that have a letter in the numerator.

In lesson 26, students first respond to a series of verbal problems in which they tell what fractions like J over 4 equals (1/4 times J).

Next, students restate fractions that have a letter and number in the numerator. What does 6B over 7 equal? 6/7 times B.

Finally, students work problems that have fractions.

> a.    $\frac{5r}{10} = 2$      b.    $\frac{h}{11} = 8$

h. Look at problem A: 5R over 10 = 2.
- Read the side with the letter. (Signal.) *5R over 10.*
- R is multiplied by what fraction? (Signal.) *5/10.*
- What's the reciprocal of 5/10? (Signal.) *10/5.*
- (Repeat step h until firm.)

i. Multiply both sides by 10/5 and write what R equals. Pencils down when you're finished.
(Observe students and give feedback.)
• (Write on the board:) [26:4D]

a. $\left(\frac{10}{5}\right) \frac{5r}{10} = 2 \left(\frac{\cancel{10}^{\,2}}{\cancel{5}}\right)$

$\boxed{r = 4}$

• Here's what you should have: R = 4.
j. Problem B: H over 11 = 8.
• Read the side with the letter. (Signal.) *H over 11.*
• H is multiplied by what fraction? (Signal.) *1/11.*
• What's the reciprocal of 1/11? (Signal.) *11.*
• (Repeat step j until firm.)
k. Solve for H. Work the problem and write what H equals. Pencils down when you're finished.
(Observe students and give feedback.)
• (Write on the board:) [26:4E]

b. $(11) \frac{h}{11} = 8\,(11)$

$\boxed{h = 88}$

• Here's what you should have: H = 88.

*Teaching note:* Make sure students respond correctly to all the verbal questions before they work each written problem. An important part of the strategy students learn is to start with the side that has a letter. Next, they figure out what to multiply by on that side. Then they multiply the other side by the same value. If students follow these steps, they will be well prepared for later algebraic problems.

In lessons 28 and 29, students work mixed sets of problems. For some, they multiply both sides by the reciprocal; for others, they add or subtract on both sides.

**TWO-STEP PROBLEMS (Lessons 30–32)**

In lessons 30–32, students learn to work two-step problems. Students learn to identify the side with the letter and the term with the letter. The problem-solving steps they follow are first to get the letter term alone on the side and then solve for the letter.

Here's the introduction from lesson 30:

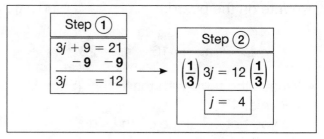

• To solve some problems with letters, you have to do more than 1 step.
b. Look at the problem in the first box: 3J + 9 = 21.
We solve this problem for J.
• What do we solve for? (Signal.) *J.*
• Everybody, read the side with J. (Signal.) *3J + 9.*
c. The **term** with the letter is 3J.
• What's the **term** with the letter? (Signal.) *3J.*
d Listen: What's the **side** with the letter? (Signal.) *3J + 9.*
• What's the term with the letter? (Signal.) *3J.*
• (Repeat step d until firm.)
e. Remember, the side is everything on that side. The term with the letter is just 3J.
• So the first thing you do is get 3J alone on the side.
• What's the first thing you do? (Signal.) *Get 3J alone on the side.*
• That means you have to get rid of + 9.
• How do you do that? (Signal.) *Subtract 9.*
Yes, subtract 9 from each side.
f. Touch the equation you get after you subtract 9 from each side. √
• Read the equation. (Signal.) *3J = 12.*
• What do you do to solve for J? (Signal.) *Multiply by 1/3.*
Yes, multiply both sides by 1/3.
g. The next box shows that step.
• So the simple equation is J = 4.
• Remember, first get the letter term alone on the side, then solve for the letter.

Here's the part of the exercise that directs students to work the first workbook problem.

- (Write on the board:)  [30:2A]

> **a.**   $16 + 4r = 28$

- What's the term with the letter? (Signal.) *4R.* Yes, 4R.
- So you get 4R alone on the side.
- What do you do to get 4R alone? (Signal.) *Subtract 16.*
- Do it. Subtract 16 from both sides and write what 4R equals. Pencils down when you've done that much. √
- (Write to show:)  [30:2B]

> **a.**   $16 + 4r = 28$
> $\underline{-16 \qquad -16}$
> $4r = 12$

- Here's what you should have: 4R = 12.
- Now figure out what R equals, and write the simple equation. Pencils down when you're finished.
  (Observe students and give feedback.)
- (Write to show:)  [30:2C]

> **a.**   $16 + 4r = 28$
> $\underline{-16 \qquad -16}$
> $\left(\frac{1}{4}\right) 4r = 12 \left(\frac{1}{4}\right)$
> $r = 3$

- Here's what you should have.
- You multiplied both sides by 1/4. R = 3.

In lessons 33 and 34, students work mixed sets of problems. Some require one-step solutions; others require two-step solutions.

## SUBSTITUTION (Lessons 37–39)

Lessons 37–39 introduce the concept that a value may be substituted for a letter. This sequence builds on what students have learned about letter terms. In lesson 37, students learn that they can't solve an equation that has two letters. If a number value is given for one of the letters, they can solve for the other letter.

Here's part of the introduction:

> $3q - 8m = 11$ $\qquad \boxed{m = \frac{1}{2}}$
> $3q - 8\left(\frac{1}{2}\right) = 11$
> $3q - 4 \quad = 11$

- c. The box shows a problem with 2 different letters: 3Q − 8M = 11.
- d. What's the first letter term? (Signal.) *3Q.*
- What's the other letter term? (Signal.) *− 8M.*
- (Repeat step d until firm.)
- You can't solve this problem unless you have a number value for 1 of the letters. If you know the value for Q, you can figure out M. If you know the value for M, you can figure out Q.
- Remember, if there are 2 letters, you have to get a number value for 1 of the letters before you can solve the problem.

Students also learn that a simple equation (M = 1/2) tells what letter they will replace with a number in the equation with two letters.

The exercise gives students practice in applying this procedure to problems that have more than one letter.

> **a.** $m - 15 = 3r$  $\qquad \boxed{r = 8}$

- Each item shows an equation with 2 letters. It also shows a simple equation that gives a number value for 1 of the letters. The simple equation is boxed.

b. Item A. Read the equation with 2 letters. (Signal.) *M − 15 = 3R.*
- Read the simple equation. (Signal.) *R = 8.*
- Which letter are you going to replace with a number? (Signal.) *R.*
- Read the R term in the first equation. (Signal.) *3R.*
- Say that term with a number for R. (Signal.) *3 (8).*
- (Repeat step b until firm.)

e. You're going to write the first equation so it has only 1 letter term. Copy both equations. Below, rewrite the first equation with 1 of the letters replaced with a number. Remember to show the parentheses for times. Pencils down when you've done that much. √
- (Write on the board:)  [37:1A]

> **a.**   $m − 15 = 3r$   $\boxed{r = 8}$
> $m − 15 = 3 (8)$

- Here's what you should have.
- The next thing you do is write the equation without parentheses. Do it. Pencils down when you've done that much. √
- (Write to show:)  [37:1B]

> **a.**   $m − 15 = 3r$   $\boxed{r = 8}$
> $m − 15 = 3 (8)$
> $\boxed{m − 15 = 24}$

- Here's what you should have.
- Now you have a problem you know how to solve. You'll do that later. Leave space for your work.

*Teaching note:* Students have learned that parentheses mean times and that a term like 3R means 3 times R.

These are the steps for solving the problem:

(1) copy the problem;

(2) replace one of the letters with a number value;

(3) remove parentheses;

(4) solve for the letter.

In this exercise, students follow the steps through removing the parentheses. They perform these steps with all the problems. Later, they solve the problems as part of their Independent Work. Make sure students have the correct equations.

### LIKE TERMS (Lessons 42–59)

In lessons 42 and 43, students work with expressions that have like terms. At this point in the program, students have learned to combine signed numbers. The introduction to like terms shows them they can also combine like letter terms.

Here's the first example they work:

$$−4m + 6m + 11m$$

The teaching box makes the point that the combined value will have M and that the procedure for combining is the same as that used for signed numbers.

Students next learn to identify which terms are like terms. Here's part of the introduction from lesson 42:

> $− 7k + \mathbf{6b} − 14 + 2p − \mathbf{4b}$
> $− 7k + \mathbf{2b} − 14 + 2p$

- Touch the first term. √
- What's the first term? (Signal.) − *7K.*
- What's the next term? (Signal.) + *6B.*
  You can't combine those terms, because 1 is a K term and the other is a B term.
- What's the next term? (Signal.) − *14.*
  That's a number term. You can't combine that term with either the K term **or** the B term.
- What's the next term? (Signal.) + *2P.*
- Raise your hand when you know if you can combine + 2P with another term. √
- Can you combine 2P with any other term? (Signal.) *No.*
  Right, it's a P term.
- What's the last term? (Signal.) − *4B.*
- Is there another B term in this problem? (Signal.) *Yes.*
- Touch that term. √
- What is that term? (Signal.) + *6B.*
- Both of them are B terms, so you can combine + 6B and − 4B. What do you get when you combine them? (Signal.) + *2B.*
- (Repeat until firm.)
  f. Yes, plus 2B. You can see the expression rewritten below with the B terms combined.

Students then simplify the following set of expressions.

| | |
|---|---|
| a. | $4p - 2m + 3 - 6p - 7$ |
| b. | $4 - 3q + 7v - 6q - 3$ |
| c. | $-k + 5 + 2k + 5k - 11$ |
| d. | $-12j + 10 + 8j - j - 9h$ |

*Teaching note:* The presentation is structured so that students identify terms that can be combined. Students write the combined terms below. Some students may show progressive expressions each with a new combined term. Tell them they can work the problem more quickly by showing all the combined terms in a single expression.

In lesson 44, students solve equations that have like terms. The procedure they follow is to combine like terms on one side, copy the rest of the equation, and solve for the unknown.

Here's how the first problem is presented. The problem is

$$5r - 2 + 3 + 8r - 6r = 3.$$

e. Look at problem A. One side has R terms and number terms.
- Can you combine R terms? (Signal.) *Yes.*
- Can you combine number terms? (Signal.) *Yes.*
- You're going to show a rewritten equation below. Remember, show the like terms combined on 1 side. Then copy the rest of the equation. Do that much. Then stop. √
- (Write on the board:)                    [44:1A]

| | |
|---|---|
| a. | $5r - 2 + 3 + 8r - 6r = 3$ |
| | $7r + 1 = 3$ |

- Here's what you should have.
- Work the problem and write what R equals. Pencils down when you're finished.
  (Observe students and give feedback.)
- (Write to show:)                    [44:1B]

$$
\begin{aligned}
\text{a.} \quad 5r - 2 + 3 + 8r - 6r &= 3 \\
7r + 1 &= 3 \\
-1 \quad &-1 \\
\hline
\left(\tfrac{1}{7}\right) 7r &= 2 \left(\tfrac{1}{7}\right) \\
r &= \tfrac{2}{7}
\end{aligned}
$$

- Here's what you should have.
- You subtract 1 from both sides.
- Then you multiply both sides by 1/7. R is 2/7.

*Teaching note:* If students are firm on the procedures for combining like terms in expressions, this exercise should not be difficult for them.

In lessons 48 and 49, the program introduces problems that have letter terms on both sides of the equation and problems that have like terms on both sides of the equation. Students follow the procedure of first combining like terms on each side of the equation, then possibly adding or subtracting a letter term on both sides.

Here's how the first problem in lesson 49 is structured:

$$9k + 6 - 4k = 6k + k$$

b. The box shows a problem: 9K + 6 − 4K = 6K + K.
- This problem has letter terms and a number term.
- The first thing you do is combine like terms.
- What's the first thing you do? (Signal.) *Combine like terms.*
- You combine like terms on each side.

c. Read everything on the side with the number term. (Signal.) *9K + 6 − 4K.*
- Can you combine number terms on that side? (Signal.) *No.*
- Can you combine letter terms on that side? (Signal.) *Yes.*
- So the first thing you do is combine like terms on that side. Say the problem you'll work for combining the like terms. (Signal.) *9K − 4K.*
- (Repeat step c until firm.)

d. Look at the other side. Are there like terms on that side? (Signal.) *Yes.*
- Say the problem you'll work for combining the like terms. (Signal.) *6K + K.*
- Copy the equation. Below, write the equation you get when you combine the like terms on each side. Stop when you've done that much.
  (Observe students and give feedback.)
- (Write on the board:) [49:5A]

$$9k + 6 - 4k = 6k + k$$
$$5k + 6 = 7k$$

- Here's what you should have.

- This is a problem you can work. You eliminate 1 of the letter terms so you have 1 letter term on 1 side and 1 number term on the other side.
- Raise your hand when you know how you change both sides. √
- Everybody, how do you change both sides? (Signal.) *Subtract 5K.*
- Write the equation with a letter term on 1 side and the number term on the other side. Then stop.
  (Observe students and give feedback.)
- (Write to show:) [49:5B]

$$9k + 6 - 4k = 6k + k$$
$$5k + 6 = 7k$$
$$\underline{-5k \qquad -5k}$$
$$6 = 2k$$

- Here's what you should have.

*Teaching note:* The solution to these problems involves many steps, which means there are more opportunities for students to make errors. The critical details students must attend to are the sequence of problem-solving steps:

(1) Copy the equation.

(2) Rewrite the equation with like terms combined.

(3) Operate on the equation and rewrite it with a single letter term on one side of the equation and a single number term on the other.

(4) Multiply by the reciprocal to solve for the letter.

(5) Write the simple equation for what the letter equals.

The only new step is step 3; however, it involves several possible operations. If students are firm on what has been taught in the program, they should be able to follow this sequence of steps reliably; however, they need practice before they become facile in executing the procedures.

If students have difficulties with particular problems, repeat those problems at a later time, possibly before beginning the next lesson.

In lessons 50 through 53, students work different configurations of problems that require combining like terms. The goal of these exercises is to give students ample practice in solving problems with like terms before they learn the next problem type.

In lessons 54–57, another step is added. Students first combine like terms, then substitute a number value in one of the letter terms, remove parentheses, and solve for the remaining letter. After they remove parentheses, the resulting equation is one that students know how to solve.

Here's the structure for the first problem students work:

| **a.** $5t + 2t = 4k$ | $t = 8$ |
|---|---|

**b.** Problem A: 5T + 2T = 4K. T = 8.
- Copy both equations. Follow the steps for working the problem: Combine like terms. Replace a letter with a number. Remove parentheses. Pencils down when you have an equation with the parentheses removed.
  **(Observe students and give feedback.)**
- (Write on the board:)                    [54:4A]

| **a.** | $5t + 2t = 4k$ | $t = 8$ |
|---|---|---|
| | $7t \quad = 4k$ | |
| | $7(8) \ = 4k$ | |
| | $56 \quad = 4k$ | |

- Here's what you should have so far. When you combine the T terms, you get 7T = 4K.
- You replace T with 8. When you remove the parentheses, you have 56 = 4K. You'll solve for K later.

*Teaching note:* Students are familiar with all the components. The only new aspect is that students combine like terms before they substitute.

In lessons 56 and 57, students work problems that have terms with two letters (5HG). They substitute for one of the letters and write parentheses to show that all the components are multiplied.

Here's the introduction from lesson 56:

| $5hg - g = 27$ | $h = 2$ |
|---|---|
| $5(2)g - g = 27$ | |
| $10g - g = 27$ | |
| $9g \quad = 27$ | |

**b.** The box shows a new kind of problem: 5HG – G = 27. 1 of the terms has more than 1 letter: That's 5HG.
- The first equation has 2 different letter terms: HG and G. 5HG is 5 times H times G.
- What is 5HG? (Signal.) *5 times H times G.*
- What is 9PQ? (Signal.) *9 times P times Q.*
- What is 1 half RS? (Signal.) *1/2 times R times S.*
- What is 2G? (Signal.) *2 times G.*
- You can't combine like terms as the first step because this problem does not have like terms. So the first thing you do is replace H with 2.

**c.** The next equation shows what you get when you replace H with 2. Everybody, read the equation. (Signal.) *5 times 2 times G minus G = 27.*
- Now you remove the parentheses. Everybody, read the equation with the parentheses removed. (Signal.) *10G – G = 27.*
- Does this problem have like terms? (Signal.) *Yes.*

**d.** The last equation shows the G terms combined. Read the equation. (Signal.) *9G = 27.*
- Now you can figure out G.

**e.** Remember, a term that has 2 letters cannot be combined with a term that has only 1 of those letters.

**Teaching note:** The two difficulties students often have are (1) not clearly showing that the values are multiplied and (2) trying to combine terms that have two letters with terms that have one of those letters. Remind them that letter terms that are combined must have the same letters. Remind them that the letters within a term are multiplied.

The last exercise in the basic algebra sequence, in lesson 59, presents a mixed set of problems that vary in whether the letter terms have single letters or double letters, whether the problem requires substitution, and whether there are like terms on one side or both sides of the equation.

Here's the set of problems from lesson 59:

a. $2w + \frac{1}{2}tw = 8t - 2t$    $\boxed{w = 6}$

b. $11 - 2r + 5r - 10 = 10 + 3$

c. $20n + 4 = 3nk - 2k$    $\boxed{n = 2}$

d. $2 = \frac{1}{4} + 8h + \frac{3}{4} - 7h$

# Algebra Translation (Lessons 22–114)

Starting at lesson 22 and continuing intermittently throughout the program, students learn how to translate various problems into letter equations that they solve algebraically.

The first type is classification problems (lesson 22). These problems involve two subsets that make up a larger set:

> *There are 28 students in a classroom. 12 are boys. How many are girls?*

The next type presents simple comparisons (lesson 34) that involve addition-subtraction or multiplication:

> *The dog weighs 12 pounds more than the cat. The cat weighs 9 pounds. How much does the dog weigh?*

> *The cat weighs 3/5 as much as the dog. The dog weighs 24 pounds. How much does the cat weigh?*

Percent word problems are introduced in lesson 62:

> *56% of the workers are members of the union. There are 1300 workers. How many are in the union?*

Variations in which parts and the whole are relevant to the solution begin in lesson 68:

> *56% of the workers were members of the union. 200 workers are not in the union. How many workers are there in all?*

More complex comparisons are introduced in lesson 77:

> *The distance to the lake is 20 miles less than 3/5 the distance to the river. The lake is 15 miles away. What is the distance to the river?*

In lessons 71–73, students work problems that ask the questions "What fraction . . . ?" and "What percent . . . ?"

> *Fifteen of 60 plates are cracked. What percent of the plates are cracked? What fraction of the plates are cracked?*

More complex problems that ask "What fraction . . . ?" and "What percent . . . ?" are introduced in lessons 102 and 103.

Problems involving greater than and less than are introduced in lesson 85:

> *The distance to the lodge is more than 7/8 of the distance to the farm. The distance to the farm is 12.8 miles. What's the distance to the lodge?*

Problems involving the combination signs ≤ and ≥ begin in lesson 93:

> *The cost of the coat is at least 2.5 times the cost of the shoes. The coat costs $100. What's the cost of the shoes?*

Problems involving increase or decrease are introduced in lesson 106:

> *The coat is on sale at 20% discount. By buying the coat on sale, you save $18. What's the original price of the coat?*

## CLASSIFICATION PROBLEMS (Lessons 22–27)

Students write a general equation form that expresses the relationship between the larger set and the subsets:

$$A = B + C.$$

Here's part of the introduction from lesson 22:

| children = boys + girls |
|---|
| $c$ = $b$ + $g$ |

| mixture = sand + gravel |
|---|
| $m$ = $s$ + $g$ |

| **a.** birds = red birds + black birds |
|---|

b. To work some problems, you need to write letters that stand for names.
- The first box shows an equation with names: Children equals boys plus girls.
- The statement is true. If you add the number for boys and the number for girls, you'll have the number for children.
- The **whole** group is children. The **parts** that are added are boys and girls.
- What's the whole? (Signal.) *Children.*
- What are the parts that are added? (Signal.) *Boys and girls.*
- Below, the equation is shown with letters: C = B + G.

c. The next box has another equation with words: Mixture equals sand plus gravel.
- What's the whole? (Signal.) *Mixture.*
- What are the parts that are added? (Signal.) *Sand and gravel.*
- The equation with letters is M = S + G.

d. For each item, you'll write an equation with letters.

e. Sentence A: Birds equals red birds plus black birds.
- You have to use 2 letters for red birds and black birds. Write the equation. Pencils down when you're finished. √

- (Write on the board:)  [22:3A]

| | |
|---|---|
| **a.** | $b = rb + bb$ |

- Here's what you should have: B = RB + BB.
- What's the whole? (Signal.) *Birds.*
- What are the added parts? (Signal.) *Red birds and black birds.*

Students write letter equations for 3 more items. They don't work a problem but simply do the translation of a sentence into a letter equation.

Students work entire problems in lesson 23. They write an equation with letters, an equation with 2 numbers, and a simple equation to show what the remaining letter equals. For example:

*There are 38 people at the picnic. There are 17 adults. How many children are at the picnic?*

$$
\begin{aligned}
p &= a + c \\
38 &= 17 + c \\
-17 &\quad -17 \\
\hline
21 &= c
\end{aligned}
$$

*Teaching note:* Students are often able to work these problems without writing an equation with letters. Even for students who have no trouble, the model shows them that word problems can be translated into equations with letters and that the algebraic steps they have learned apply to these equations.

Tell students that they will do a lot of translations of words to symbols in equations.

Later in the program, the *Answer Key* continues to show this type of problem solved with a letter equation. However, don't require students to write the equation if they have no trouble creating a correct addition or subtraction problem from the information the word problem provides.

## COMPARISON PROBLEMS (Lessons 31–58)

First students learn to write a letter equation for statements that present the relationship between two unknowns; for instance,

$$j \text{ is } 5 \text{ more than } p.$$

Students write $j = p + 5$.

$$m \text{ has } 12 \text{ fewer than } p.$$

Students write $m = p - 12$.

Expressing these relationships properly is important for problem types that are to come. Often students are confused about how to write the equation for items that tell about less. The reason is that problems that tell about more may be written with the number first. Four more than J is $4 + J$. If the problem says that $t$ is 6 less than $d$, some write $t = 6 - d$. This equation is not right. The introductory exercises stress that the letter always comes first after the equal sign.

Here's part of the exercise from lesson 32:

| ◆ The part that adds or subtracts always starts with a letter. | ◆ 3 more than $f$ <br> $f + 3$ | ◆ 12 less than $b$ <br> $b - 12$ |
|---|---|---|

b. For some statements, you write an equation that adds or subtracts.
- The part that adds or subtracts always starts with a letter.

c. The box shows a part that tells about adding: 3 more than F.
- Say that part. (Signal.) *3 more than F.*
- For that part, you start with the letter F and write F plus 3.
- What do you write for 3 more than F? (Signal.) *F + 3.*

d. What do you write for 72 more than P? (Signal.) *P + 72.*
- What do you write for 13 more than D? (Signal.) *D + 13.*

e. The box also shows a part that tells about subtracting: 12 less than B.
- You start with the letter and write B − 12.
- What do you write for 12 less than B? (Signal.) *B − 12.*

f. What do you write for 2 less than R? (Signal.) *R − 2.*
- What do you write for 9 less than Z? (Signal.) *Z − 9.*

In lesson 33, students translate statements that call for multiplication.

| ◆ B is 6 **times** A. <br> ◆ B = 6A |
|---|

◆ $f$ is 5 times $n$.

◆ $k$ is $\frac{2}{3}$ of $p$.

◆ $h$ is $\frac{1}{2}$ as much as $d$.

- Not all sentences that compare add or subtract. Some multiply.
- For a part that **multiplies,** you start with the **number.**

e. You can see the statement B is **6 times A.**
- Say the statement. (Signal.) *B is 6 times A.*
- You write B = **6A.**
- What do you write for B is 6 times A? (Signal.) *B = 6A.*
  That's B = 6 **times** A.

f. Next sentence: F is 5 times N.
  Say the equation. (Signal.) *F = 5N.*
- Next sentence: K is 2/3 of P.
  Say the equation. (Signal.) *K = 2 thirds P.*
- Next sentence: H is 1 half as much as D.
  Say the equation. (Signal.) *H = 1 half D.*

*Teaching note:* Following this introduction, students write equations for sentences that call for addition-subtraction or multiplication. Students practice this discrimination on the following lessons. It is important for students to become reliable on the translation. Expect some students to have difficulties because the translation for one type of sentence calls for the letter to be shown first; the other requires the number to be shown first.

In lesson 39, students work complete problems that require addition, subtraction, or multiplication. They write an equation with letters, substitute a value for one of the letters, and solve for the other letter.

a. *The cat was 7 inches shorter than the dog. If the cat was 9 inches tall, how tall was the dog?*

$$c = d - 7$$
$$9 = d - 7$$
$$\underline{+\,7 \qquad +\,7}$$
$$16 = d$$

| 16 inches |
|---|

b. Jane is $\frac{7}{5}$ the age of Sally. Jane is 56 years old. How old is Sally?

$$J = \frac{7}{5} S$$

$$\left(\frac{5}{7}\right) 56 = \frac{7}{5}S \left(\frac{5}{7}\right)$$

$$40 = S$$

$$\boxed{40 \text{ years old}}$$

**Teaching note:** If students are at mastery on working the algebra problems and at translating key sentences into letter equations, they should have no trouble with these exercises (except possibly with calculating correctly).

In lessons 40–58, students work problems that present variations in wording. Here are the major variations they learn to work:

a. **It is 20 miles farther to the town than to the forest. If it is 66 miles to the forest, how far is it to the town?**

$$t = f + 20$$
$$t = 66 + 20$$
$$t = 86$$

$$\boxed{86 \text{ miles}}$$

b. **It is $\frac{1}{5}$ as far to the town as it is to the lake. The distance to the town is 22.5 miles. How far is it to the lake?**

$$t = \frac{1}{5}l$$
$$(5)\ 22.5 = \frac{1}{5}l\ (5)$$
$$112.5 = l$$

$$\boxed{112.5 \text{ miles}}$$

Translation for these problem types is more difficult because the first sentence must be rephrased. The sentence "It is 20 miles farther to the town than to the forest" translates into "The town is 20 miles farther than the forest." Students first translate sentences before working complete problems. The simplest way to prompt the translation is to say, "Start with the first thing named, and say the sentence another way."

Following the introduction of each new type, students work mixed sets of problems that include the type that has been most recently introduced and the other types that might be confused with this type.

**FRACTION/PERCENT EXTENSIONS**

In lesson 61, students learn to translate statements of this type: 3/7 of the girls are working. For these statements, students write the terms in the same order the sentence presents the relationship:

$$\frac{3}{7}g = w.$$

This is an extension of what they have learned earlier, that if the first part of the sentence tells what something costs, is, has, or does, students write a letter for what is named followed by an equal sign. *Girls are* is expressed as G =. *3/7 of the girls are* is expressed as *3/7 G =*. The same symbols would be written for *3/7 of the girls have, 3/7 of the girls eat,* or *3/7 of the girls spend.*

Percent problems are introduced in lesson 52:

**Al's height is 120% of his dad's height. If Al is 78 inches tall, how tall is his dad?**

Students write $A = \frac{120}{100} d.$

**55% of the dogs have been fed. If 22 dogs have been fed, how many dogs are there in all?**

Students write $\frac{55}{100} d = f.$

Problems that ask about the fraction or percent are introduced in lesson 71.

**24 of the 120 voters voted yes. What percent of the voters voted yes?**

Students base the letter equation on the question.

They write $\frac{p}{100} (v) = y.$

Note that the unknown is a part of the fraction.

They learn a variation for questions that ask what fraction.

**What fraction of the boys are sleeping?**

Students write $f (b) = s.$

## PARTS OF A SET

Students earlier worked problems based on a larger set composed of two subsets. For those problems, students added or subtracted.

> *She had 28 roses. 4 were white. How many were not white?*

Students have also worked simple subset problems that involved fractions.

> *2/7 of the roses were white. If there were 28 roses, how many white roses were there?*

Beginning in lesson 68, students work problems for which they write equations for two subsets.

> *2/7 of the roses are white. There are 28 roses in all. How many are not white?*

To work this problem, students write an equation for white roses and not-white roses:

$$\frac{2}{7}r = w \qquad \frac{5}{7}r = nw$$

Prior to working this problem type, students solve two complementary part-whole equations (lessons 65 and 66). Students solve one equation and then substitute in another.

For example, students work this pair of equations:

$$\frac{1}{3}t = j \qquad \frac{2}{3}t = k \qquad \boxed{j = 5} \qquad k = ?$$

The simple equation shows that J equals 5. Students first substitute 5 for J in the first equation and solve for T:

$$\frac{1}{3}t = j$$

$$(3)\frac{1}{3}t = 5\,(3)$$

$$\boxed{t = 15}$$

Then students substitute in the other equation and solve for K:

$$\frac{2}{3}t = k$$

$$\frac{2}{3}(15) = k$$

$$\boxed{10 = k}$$

Also, before working word problems involving two complementary equations that describe a whole, students generate two equations from statements that express fraction relationships for subsets (lesson 67).

For example, the statement indicates 5/7 of the balls were yellow. Students write two equations:

$$\frac{5}{7}b = y \qquad \frac{2}{7}b = ny$$

In lesson 68, students work complete word problems that require two equations, one for each subset named in the problem.

Here's part of the introduction from lesson 68:

---

**a.** 60% of the children are boys. There are 12 boys. How many girls are there?

---

- Each problem tells about 1 name but asks about another name. So you write a letter equation for the part the problem tells about and another equation for the part the question asks about.
- b. Problem A: 60% of the children are boys. There are 12 boys. How many girls are there?
- 60% is the same as 60/100.
- Write the 2 letter equations for this problem. Then stop. √
- (Write on the board:)                    [68:1A]

**a.** $\qquad \dfrac{60}{100}c = b \qquad \dfrac{40}{100}c = g$

- Here are the 2 equations you should have. The first sentence tells you that 60 hundredths C = B.
- The other equation is 40 hundredths C = G.
- The problem gives a number for 1 of the letters. Write the simple equation for that letter. Then substitute for that letter and solve 1 of the equations. Stop when you've done that much.

(Observe students and give feedback.)

- (Write to show:) [68:1B]

a.
$$\frac{60}{100}c = b \qquad \frac{40}{100}c = g \qquad \boxed{b = 12}$$

$$\left(\frac{100}{60}\right)\frac{60}{100}c = 12\left(\frac{100}{60}\right)$$

$$\boxed{c = 20}$$

- Here's what you should have. You substituted 12 for B. You solved for C. What does C equal? (Signal.) *20.*
- Now substitute for C in the other equation. Figure out what G equals and write the answer to the problem. Pencils down when you're finished.
(Observe students and give feedback.)
- (Write to show:) [68:1C]

a.
$$\frac{60}{100}c = b \qquad \frac{40}{100}c = g \qquad \boxed{b = 12}$$

$$\left(\frac{100}{60}\right)\frac{60}{100}c = 12\left(\frac{100}{60}\right) \qquad \frac{40}{100}(20) = g$$

$$\boxed{c = 20} \qquad \qquad 8 = g$$

$$\boxed{8 \text{ girls}}$$

- Here's what you should have. If there are 12 boys, there are 8 girls.

*Teaching note:* These problems may be solved another way, by figuring out the number of children, then subtracting the number of boys. The reason the substitution method is presented it that it works for all subtypes and does not require students to learn tricky discriminations about which of the two subset equations to write.

Here's an example of the type that is not difficult with the procedure students learn, but that could be clumsy using the add-subtract procedure:

*60 percent of the children are boys. There are 8 girls. How many boys are there?*

If students write equations for both subsets as a standard procedure, they will be able to solve the full range of problems.

The following exercises (lessons 69–73) present mixed sets containing problems of these two types and a third type that refers to the number for the whole group.

## TRANSLATIONS INVOLVING MORE THAN ONE OPERATION (Lessons 74–79)

In lessons 74 and 75, students translate sentences resulting in equations that involve more than one operation.

> *Ron's weight is 7 pounds less than 4 times Bill's weight.*

Students write $R = 4B - 7$.

This translation requires students to apply what they have learned about equations that specify multiplication ($4B$) and equations that specify adding-subtracting ($- 7$).

Here's part of the presentation from lesson 74:

| **Bill is** 3 years older than **2 times Harry's age.** |
|---|
| **B =**            **2H**      + 3. |

b. You've learned how to write equations from sentences that compare 2 things.
- The box has a more complicated sentence that compares 2 things: Bill is 3 years older than 2 times Harry's age.
- Everybody, say that sentence. (Signal.) *Bill is 3 years older than 2 times Harry's age.*
- For the words **Bill is,** we write **B equals.**
- Whose age is compared with Bill's age? (Signal.) *Harry's.*
- For the words **2 times Harry's age,** we write **2H.**
- For the words **3 years older than,** we write **plus 3.**
- The complete equation for "Bill is 3 years older than 2 times Harry's age" is B = 2H + 3.

---

a. **Ron's weight is** 7 pounds less than **4 times Bill's weight.**

b. **Car M goes** 50 feet more than $\frac{3}{4}$ the distance car N goes.

c. **Factory A produced** 300 fewer cartons than **120% of the cartons produced by factory B.**

c. Sentences A through F are just like the one we just did.

- Each sentence tells about multiplying and about adding or subtracting.
- In the first 3 sentences, the part that tells about multiplying is shown in red.

d. Sentence A: Ron's weight is 7 pounds less than 4 times Bill's weight.

- What's the part that tells about multiplying? (Signal.) *Ron's weight is 4 times Bill's weight.*
- What do you write for that part? (Signal.) *R = 4B.*
- What's the part that tells about adding or subtracting? (Signal.) *7 pounds less than.*
- What do you write for that part? (Signal.) *− 7.*

e. Sentence B: Car M goes 50 feet more than 3/4 the distance car N goes.

- What's the part that tells about multiplying? (Signal.) *Car M goes 3/4 the distance car N goes.*
- What do you write for that part? (Signal.) *M = 3 fourths N.*
- What's the part that tells about adding or subtracting? (Signal.) *50 feet more than.*
- What do you write for that part? (Signal.) *+ 50.*

f. Sentence C: Factory A produced 300 fewer cartons than 120% of the cartons produced by factory B.

- What's the part that tells about multiplying? (Signal.) *Factory A produced 120% of the cartons produced by factory B.*
- What do you write for that part? (Signal.) *A = 120 hundredths B.*
- What's the part that tells about adding or subtracting? (Signal.) *300 fewer cartons than.*
- What do you write for that part? (Signal.) *− 300.*

After going through a series of 6 verbal examples, students write equations for all the sentences they responded to verbally.

*Teaching note:* Make sure students are firm on translating sentences before Lesson 76 because in that lesson, they will work entire problems.

The problems students work in lessons 76 and 77 have sentences like the three above. All translate into an equation that starts with a letter and an equal sign.

In lesson 78, sentences for more complicated problems are introduced:

**Two times the length of the train is 50 feet more than 1/2 the length of the bridge.**

Students write $2t = \frac{1}{2}b + 50.$

In lesson 79, students work complete problems that require equations of the type practiced in lesson 78.

*Teaching note:* Although the wording of the problems describes complicated relationships, the process of writing the equation with letters is fairly manageable and students are able to master it.

Mastery of these problems provides students with evidence that they are able to work problems that many higher-performing students would have trouble working. Remind students that they should feel proud because the problems require substitution, expressing relationships of three operations, and performing multi-step algebraic solutions. Good performance on working these problems is strong evidence of substantial learning.

### INEQUALITIES (Lessons 79–96)

In lessons 79 and 80, students apply what they have learned about solving equations to solving inequalities. Students perform operations on both sides and solve for a simple inequality that has a letter and a number.

Students first review the conventions for the inequality signs. They practice saying simple inequalities starting with the left value and starting with the right.

For example: $7 > 3.$

Students first say, "Seven is greater than 3."

Then they say, "Three is less than 7."

The relationship is important for some of the problems students will work later.

Next, students work problems that have inequality signs. Here's part of the introduction in lesson 79:

e. You can work problems with an inequality sign the same way you work problems with an equal sign.

f. (Teacher reference:)

$$4k - 3 > 17$$
$$\underline{\phantom{4k} + 3 \phantom{>} + 3}$$
$$4k \phantom{-3} > 20$$

$$\left(\tfrac{1}{4}\right) 4k \phantom{--} > 20 \left(\tfrac{1}{4}\right)$$

$$\boxed{k \phantom{--} > 5}$$

- I'll read the problem in the next box: 4K − 3 is greater than 17.
- Everybody, say the statement that starts with 4K. (Signal.) *4K − 3 is greater than 17.*
- We work the problem the same way we work a problem with an equal sign: We solve for K. So the first thing we do is get the K term alone on 1 side of the inequality sign and the number term on the other side.
- What do we do to get the K term alone? (Signal.) *Add 3.*
- You can see the inequality we get when we add 3 to both sides: 4K is greater than 20.
- Everybody, say the statement that starts with 4K. (Signal.) *4K is greater than 20.*
- Now we multiply both sides by 1/4 and we end up with the inequality that tells about K: K is greater than 5.
- That means any number that is more than 5 could be K.

In lesson 82, students translate sentences into equations or inequalities. Students learn that if the sentence tells how many more or how many less, they write an equation. If there is no number for how many more or less, they write an inequality.

*4 times T is less than K.* Students write **4t < k.**

*4 times T is 3 less than K.* Students write **4t = k − 3.**

In lesson 83, students write an inequality statement or equation, substitute for one of the letters, and solve for the other letter.

In lessons 86 and 87, students write and solve inequalities for word problems.

For example:

*Ann went $\tfrac{2}{3}$ the distance to the lake and then went another 15 miles, but she was still less than half the distance to Tinkertown. It was 182 miles to Tinkertown. What was the distance to the lake?*

Students write an inequality for the first sentence:

$$\tfrac{2}{3}L + 15 < \tfrac{1}{2}T$$

Students substitute for T and solve for L:

$$L < 114$$

Students answer the question the problem asks:

$$\boxed{\textbf{less than 114 miles}}$$

In lessons 88–96, students learn to translate problems that call for a combination symbol: ≤, ≥.

The development is the same as that for inequalities. Students first learn to write statements from sentences and discriminate between problems that call for combination symbols and those that call for a simple inequality.

The last instruction on problem types occurs in lessons 91–93. Students work with sentences that have the following expressions: "at least," "at most," "no more than," "no less than," "a minimum of," and "not exceeding." In lesson 92, students write statements for sentences. Here's the part of the exercise that follows the introduction of some new wording:

| | |
|---|---|
| **a.** $4j$ is at least 60. | **d.** Shoes cost at least $45. |
| **b.** $p$ is at most 54. | **e.** Her savings did not exceed $5 a week. |
| **c.** The cost of the hats cannot exceed $12. | **f.** Roberta had a minimum of 3 appointments. |

e. You're going to write a statement for each sentence.

f. Sentence A: 4J is at least 60.
- Can 4J be 60? (Signal.) *Yes.*
- Can 4J be more than 60? (Signal.) *Yes.*
- Write the statement. √
- (Write on the board:)                    [92:2A]

| **a.**    $4j \geq 60$ |
|---|

- Here's what you should have.
- Everybody, read the statement.
  (Signal.) *4J is greater than or equal to 60.*
g. Sentence B: P is at most 54.
- Can P be 54? (Signal.) *Yes.*
- Can P be more than 54? (Signal.) *No.*
  P can be no more than 54.
- Write the statement. √
- (Write on the board:)                    [92:2B]

| **b.**    $p \leq 54$ |
|---|

- Here's what you should have.
- Everybody, read the statement.
  (Signal.) *P is less than or equal to 54.*
h. Sentence C: The cost of the hats cannot exceed $12.
- Can the hats cost $12? (Signal.) *Yes.*
- Can the hats cost more than $12? (Signal.) *No.*
- Write the statement. √
- (Write on the board:)                    [92:2C]

| **c.**    $h \leq 12$ |
|---|

- Here's what you should have.
- Everybody, read the statement.
  (Signal.) *H is less than or equal to 12.*
i. Sentence D: Shoes cost at least $45.
- Write the statement. √
- (Write on the board:)                    [92:2D]

| **d.**    $s \geq 45$ |
|---|

- Here's what you should have.
- Everybody, read the statement.
  (Signal.) *S is greater than or equal to 45.*

j. Sentence E: Her savings did not exceed $5 a week.
- Write the statement. √
- (Write on the board:)                    [92:2E]

| **e.**    $s \leq 5$ |
|---|

- Here's what you should have.
- Everybody, read the statement.
  (Signal.) *S is less than or equal to 5.*
k. Sentence F: Roberta had a minimum of 3 appointments.
- Write the statement. √
- (Write on the board:)                    [92:2F]

| **f.**    $R \geq 3$ |
|---|

- Here's what you should have.
- Everybody, read the statement.
  (Signal.) *R is greater than or equal to 3.*

*Teaching note:* The first four items are prompted with questions. The last two are not prompted.

The question routine provides a model for what to ask students who make errors. For example, students have trouble with a statement: The elephant weighed at least 3 tons.

If students write $e \leq 3$, tell them to read their statement. Then ask: Can an elephant weigh 3 tons? Can an elephant weigh more than 3 tons? An elephant can weigh more than 3 tons, so the correct symbol is greater than or equal to.

The last exercises involving inequalities, combination symbols, and equations (lessons 94–96) present a mix of problem types that students have worked.

**INCREASE/DECREASE (Lessons 106–114)**
This sequence provides students with strategies for working problems that refer to a percentage increase or decrease. The sequence introduces problems that refer to markup and discount, profit and loss. The basic strategy presented is to express the increase or decrease as a percent of the original amount. At this point in the program, students have the algebra skills required to substitute and solve the equations. What they learn is how to translate the word problems into equations with letters.

Lesson 106 introduces the basic problem type. This type presents the original amount and the percent increase or decrease. Students solve for the amount of the increase or decrease.

b. Some problems involve a percent **increase** or a percent **decrease.** The increase or decrease is a percent of the **Original Amount.** For these problems, you write an equation that tells what the increase or decrease equals.

• (Teacher reference:)

| | |
|---|---|
| $I = \dfrac{20}{100}\,OA$ | $D = \dfrac{20}{100}\,OA$ |
| • What's a 20% increase for an original amount of $40? $$I = \frac{2\cancel{0}}{1\cancel{0}\cancel{0}}\,(4\cancel{0})$$ $$I = 8$$ $\boxed{\$\,8}$ | • If the original weight is 140 pounds, what's a 20% decrease? $$D = \frac{2\cancel{0}}{1\cancel{0}\cancel{0}}\,(14\cancel{0})$$ $$D = 28$$ $\boxed{28 \text{ pounds}}$ |

c. You can see the equation for a 20% increase. I equals 20/100 times OA. I is the increase. OA is the original amount.
• Let's say you have to figure out a 20% increase for an original amount of $40.
• The equation is I = 20/100 times 40. You solve for I. The increase is $8.
d. The equation for the decrease works the same way. The next box shows the equation for a 20% decrease: D = 20% of the original amount.
• If the original amount is 140 pounds, the equation is D = 20/100 times 140, so the decrease is 28 pounds.

Following the introduction, students work basic problems that ask about the amount of the increase, the amount of the decrease, or the original amount.

Here's an example of each problem type presented in lesson 106:

a. Mark had a collection of rocks. He started with 150 rocks. He collected more rocks and increased his collection by 30%. How many more rocks did he collect?

c. The amount of money Karen had in her bank account increased by 5%. The increase was $25. How much was in the account before the increase?

d. In March, Sarah's income decreased by 8%. Before March, she earned $1200 a month. How much less did she earn in March?

In lesson 107, wording variations are presented. The problems refer to profit or loss, markup or discount, interest, and the amount saved by purchasing an item at a discount.

In lesson 108, students work problems that refer to the new amount that follows the increase or decrease:

*Packages of pencils are on sale at a 20% discount. The original price was $3.00 per package. What's the sale price?*

For problems that ask about the new amount, students write an equation that expresses the new amount as a percent of the original amount. The original amount is always 100%.

For an increase, students add to 100% to find the new percentage, and for decrease, they subtract from 100%. For the example above, students subtract 20% from 100% to find the percentage for the new amount, then substitute for the original amount, and solve for the new amount.

Here's the work from the problem above:

$$NA = \frac{80}{100}\,(OA)$$

$$NA = \frac{80}{100}\,(3.00)$$

$$NA = 2.40$$

$$\boxed{\$2.40}$$

In lesson 112, students work the last problem type, which asks about the percent increase or decrease.

| $\frac{p}{100}(OA) = I$ | $\frac{p}{100}(OA) = D$ |
|---|---|
| ◆ What percent is the increase? | ◆ What percent is the decrease? |

b. These problems ask about the percent of the increase or decrease. To work them, you write an equation for **what percent.** Remember, you write P over 100 for what percent.

c. You can see an equation for the increase and one for the decrease. Both start with P over 100. The equations relate the increase and the decrease to the original amount. If the problem asks what percent is the increase, that means what percent of the **original amount** is the increase.

d. Once more: If the problem asks what percent is the increase, what does that mean? (Signal.) *What percent of the original amount is the increase.*

• If the problem asks what percent is the decrease, what does that mean? (Signal.) *What percent of the original amount is the decrease.*

• (Repeat step d until firm.)

*Teaching note:* Students should be firm on the translation before they work problems. Expect them to have trouble saying the equations because the equations do not have parts in the same order as the sequence of words in the problem. The task is made somewhat easier if students first answer the question about whether the problem describes an increase or a decrease.

The final sets of problems students work either ask about a percent increase-decrease or tell about a percent increase-decrease.

# Coordinate System (Lessons 28–118)

This track first teaches about points on the coordinate system, then about function tables, then straight-line equations through zero ($y = mx$), and straight-line equations that do not go through zero ($y = mx + b$). A later extension shows students how to work problems that present similar triangles formed by parallel lines on the coordinate system.

## INTRODUCTION TO THE COORDINATE SYSTEM

In lessons 28 and 29, students learn how to identify the $x$ and $y$ values of points on the coordinate system. In lessons 30 and 31, students make points from information about the $x$ and $y$ values.

Here's the exercise for the second day of the introduction:

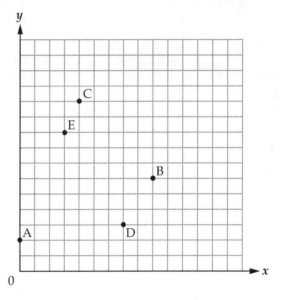

b. You're going to write the X and Y values that tell how to reach different points on the coordinate system.

• Point to show the direction for the X value. (Students point right.) √

• Point to show the direction for the Y value. (Students point up.) √

c. When you find both the X and Y values, point to show the direction you go first. (Students point right.) √
- Is that X or Y? (Signal.) *X*.
- Point to show the direction you go next. (Students point up.) √
- Yes, the X value tells how many places to the right of zero. The Y value tells how many places up.

d. Touch point A. √
- How many places to the right of zero is that point? (Signal.) *Zero places.*
- So X is zero.
- Write that equation and the equation for Y. √
- Check your work.
- (Write on the board:)         [29:4A]

| | | |
|---|---|---|
| **A** | $x = 0,$ | $y = 2$ |

- Here's what you should have for point A: X = 0, Y = 2.

e. Now write the X and Y equations for the rest of the points. Pencils down when you're finished.
(Observe students and give feedback.)
- Check your work.
- (Write to show:)         [29:4B]

| | | |
|---|---|---|
| **A** | $x = 0,$ | $y = 2$ |
| **B** | $x = 9,$ | $y = 6$ |
| **C** | $x = 4,$ | $y = 11$ |
| **D** | $x = 7,$ | $y = 3$ |
| **E** | $x = 3,$ | $y = 9$ |

f. Read both equations for each point.
- Point B. (Signal.) *X = 9, Y = 6.*
  Point C. (Signal.) *X = 4, Y = 11.*
  Point D. (Signal.) *X = 7, Y = 3.*
  Point E. (Signal.) *X = 3, Y = 9.*

The coordinate system in the early exercises does not show numbers on the $x$ and $y$ axes. The purpose of this convention is to ensure that students understand how to find points without referring to numbers on the axes. This skill is particularly important later in the sequence when students identify the slope of lines. If students are not familiar with how to operate on the coordinate system without referring to the numbers on the axes, they will have serious problems in understanding how the slope is computed. They will assume it has to do with the numbers on the axes.

In lessons 32 and 33, students connect a series of points to draw a straight line that goes through zero. They then complete a simple table that describes points shown on the line.

| | x | y |
|---|---|---|
| **A** | | 4 |
| **B** | | 2 |
| **C** | | 3 |
| **D** | | 5 |

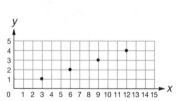

- This coordinate system has lots of points lined up.

b. Use your ruler. Very carefully draw a line through the points. The line you draw must go through the point for zero on your coordinate system. Pencils down when you're finished. √

c. You're going to complete descriptions for some points on the line. The table shows the **Y values** for the points. For each Y value, you'll write the correct X value.

d. Touch the first value for Y in the table. √

• The first value for Y is 4. Look at the coordinate system and go to 4 on the Y axis. Then go straight across to the line. Write the capital letter A above the point. √

• Now you're going to figure out the X value for point A. Go straight down to the X axis. The number on the X axis is the number for X.

• Everybody, what's the X value for point A? (Signal.) *12.*

• Write the X value for point A. Pencils down when you're finished. √

• (Write on the board:)                    [33:5A]

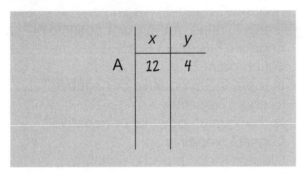

|   | x | y |
|---|---|---|
| A | 12 | 4 |

• Here's what you should have for point A: X is 12, Y is 4.

*Teaching note:* The numbers are given for the *x* and *y* axes. Students refer to these numbers when figuring out each *x* value.

In the preceding lesson, students did a similar exercise in which the *x* values were given. Students labeled specified points and then completed the table to show the *y* value for each point.

In lessons 35 and 36, the table gives information about two points. Students plot and label the points. Students draw a line connecting the points and going through zero. They then plot the other points from information the table gives (which is either the *x* or the *y* value for each point) and complete the table.

Here's the table from lesson 35:

|   | x | y |
|---|---|---|
| A | 2 | 6 |
| B | 4 | 12 |
| C |   | 9 |
| D | 1 |   |
| E | 5 |   |

The exercise in lesson 39 introduces the conventional notation for the coordinates of points (*x, y*).

In lessons 40 and 41, students label the *x* and *y* axes and write coordinates for points. Students also plot points for specified coordinates. Here's the first part of the exercise from lesson 40:

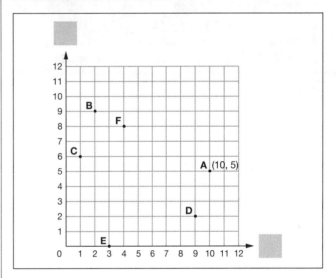

• The X axis and the Y axis are not labeled on the coordinate system.

• Label each axis. Write X in one box and Y in the other. √

• Everybody, which axis goes up and down? (Signal.) *Y.*

• Which axis goes from side to side? (Signal.) *X.*

• Make sure you have them labeled correctly.

b. You're going to write descriptions for points on this coordinate system. The description for A is already written. Remember, the numbers are in parentheses.

• The first number tells about which letter? (Signal.) *X.*

- The other number tells about Y, and there's a comma between the numbers.
c. Write descriptions for the rest of the points. Write them next to the points. Pencils down when you're finished. (Observe students and give feedback.)
- (Write on the board:) [40:4A]

> **B** (2, 9)  **D** (9, 2)  **F** (4, 8)
> **C** (1, 6)  **E** (3, 0)

- Here are the descriptions you should have for the points: B (2 comma 9); C (1, 6); D (9, 2); E (3, 0); F (4, 8).

> *Teaching note:* When you describe points, you say "comma." (Four, comma, five.) When students describe points, they are to follow this convention. Also, students are always expected to show the coordinates inside parentheses.

Lessons 43–45 introduce the four-quadrant coordinate system. Students describe points in all four quadrants.

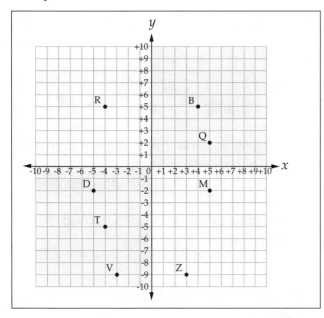

In lessons 43 and 44, students write the coordinates of different points. Students start at zero and go either left or right for *x* and either up or down for *y*, just as they would on number lines.

Here's the exercise from lesson 44:

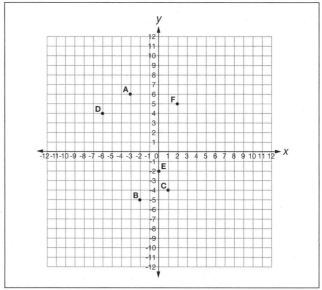

b. Touch point A. √
- Is the X value for that point positive or negative? (Signal.) *Negative.*
- Is the Y value of the point positive or negative? (Signal.) *Positive.*
c. Touch point B. √
- Is the X value positive or negative? (Signal.) *Negative.*
- Is the Y value positive or negative? (Signal.) *Negative.*
- (Repeat steps b and c until firm.)
d. You're going to write the coordinates for each point. Remember, write them next to the letter.
e. Write the coordinates for A. You don't have to write the plus sign for X or Y values that are positive. √
- What's the X value for A? (Signal.) − 3. Yes, − 3.
- What's the Y value for A? (Signal.) 6. Yes, 6.
- Say the coordinates for point A. (Signal.) − 3 comma 6.
f. Write the coordinates for the rest of the points. Remember, the coordinates tell you a route from zero. Take the route on the X axis, then go either up or down to the point. Pencils down when you're finished. (Observe students and give feedback.)
g. Check your work.

h. Point B. What's the X value? (Signal.) − 2.
i. What's the Y value? (Signal.) − 5.
j. Say the coordinates for point B. (Signal.)
   − 2 comma − 5.
• (Repeat steps h–j for points C–F.)

**Key for oral responses:**

B (− 2, − 5)    E (0, − 2)

C (1, − 4)    F (2, 5)

D (− 6, 4)

*Teaching note:* The final direction is to repeat steps h–j for points C–F. For point C, you would say, "What's the *x* value?" "What's the *y* value?" "Say the coordinates for point C." The key shows the student responses for each point.

## FUNCTION TABLES (Lessons 37–54)

The sequence first introduces functions involving addition and multiplication. Students complete tables that show *x, y,* and the function. Students plot lines based on the completed table. Here's part of the introduction from lesson 37:

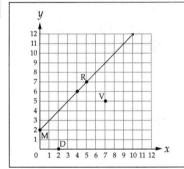

| Function | |
|---|---|
| x + 2 = | y |
| 5 + 2 | 7 |
| 0 + 2 | 2 |
| 4 + 2 | 6 |

• Each straight line on the coordinate system follows a rule. The rule tells about all points on the line. The rule is called a **function.**
• What's the rule for a straight line called? (Signal.) *A function.*
• The function tells how to figure out what Y equals. The function starts with X.
• What does the function start with? (Signal.) *X.*
• The functions you'll work with tell how to go from any X value to the corresponding Y value.

b. The first coordinate system shows a line with letters and points. The function is shown at the top of the first table. The function for the line on the coordinate system is X + 2.
• What's the function? (Signal.) *X + 2.*
• That means you can start with any X value and add 2 to figure out the corresponding Y value.
c. Listen: When the X value is 5, the Y value is 5 + 2. That's 7. Touch the point where X is 5 and Y is 7. √
• What's the letter of that point? (Signal.) *R.*
d. Listen: When the X value is zero, the Y value is zero + 2. That's 2.
• Touch the point where X = zero and Y = 2. √
• What's the letter of that point? (Signal.) *M.*

In lesson 39, students work with tables that have more than one possible function. Students test an addition rule and the multiplication rule for going from the *x* value to the *y* value.

Here's part of the exercise from lesson 39 that describes the basic procedure students follow for testing possible functions:

| | Function | | |
|---|---|---|---|
| | x (2) = | | y |
| | x + 4 = | | y |
| A | 4 | | 8 |
| B | 6 | | 10 |
| C | 3 | | |
| D | 5 | | |

e. Look at table 1.
• The X value and the Y value are shown for row A. The X value is 4. The Y value is 8. The 2 function rules show 2 different ways to get from 4 to 8. 4 times 2 is 8. And 4 plus 4 is 8.
f. Say the multiplication problem to get from 4 to 8. (Signal.) *4 times 2.*
• Say the addition problem. (Signal.) *4 + 4.*
• (Repeat step f until firm.)

g. So the functions shown are X times 2 and X + 4.

• You'll figure out which of the functions works for row B. That's the rule that works for the rest of the X values.

h. Go to row B.

• The X value is 6. The Y value is 10. Only 1 of the functions will give you the Y value of 10. Test X times 2. Then test X + 4. Raise your hand when you know which function is correct. √

• Everybody, say the correct function of X for the table. (Signal.) *X + 4.*

• That's the function that works for all the rows. Cross out the function **X times 2.** That function is wrong. √

*Teaching note:* Make sure students cross out the incorrect function. To correct mistakes, direct the students to say the multiplication function, say the addition function, and then test each function to see if it yields the *y* value given in row B. If not, students cross it out and use the other function.

Starting with lessons 40–43, students generate the two possible functions for the pair of values shown. To identify the multiplied value for a pair of values, students apply what they learned earlier to identify the fraction for a missing factor.

Here's part of the introduction from lesson 41:

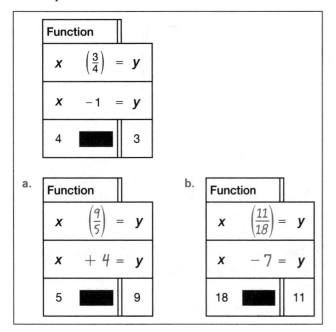

• You've worked functions that involve whole-number multiplication. You can also figure out any fraction you multiply by.

b. The sample problem shows an X value of 4 and a Y value of 3.

• To figure out the multiplication function, you ask the question: 4 times what fraction equals 3?

• Everybody, 4 times **what fraction** equals 3? (Signal.) *3/4.*

• So the multiplication function is X times 3/4.

• Everybody, what's the multiplication function? (Signal.) *X times 3/4.*

• The other function is X minus 1.

• What's the other function? (Signal.) *X − 1.*

c. Item A. What's the X value? (Signal.) *5.*

• What's the Y value? (Signal.) *9.*

• 5 times what fraction equals 9? (Signal.) *9/5.*

• So what's the multiplication function? (Signal.) *X times 9/5.*

• Raise your hand when you can say the other function. √

• What's the other function? (Signal.) *X + 4.*

• (Repeat step c until firm.)

d. Item B. What's the X value? (Signal.) *18.*

• What's the Y value? (Signal.) *11.*

• 18 times what fraction equals 11? (Signal.) *11/18.*

• So what's the multiplication function? (Signal.) *X times 11/18.*

• Raise your hand when you can say the other function. √

• What's the other function? (Signal.) *X − 7.*

• (Repeat step d until firm.)

*Teaching note:* After students orally respond to one more function, they complete five problems in which they determine two possible functions.

In lesson 44, students determine the correct function and complete a table that shows *x* and *y* values.

In lessons 45–48, students apply the equation

$$x \left( \frac{y}{x} \right) = y$$

to function tables. These tables have *x* values in the left column, and *y* values in the right column.

In lessons 52–54, students apply the same equation in reverse order:

$$y = \left( \frac{y}{x} \right) x.$$

These tables have values in the same order they appear in the equation: *y* values in the left column and *x* values in the right.

Here is the set-up for lesson 53:

| | Function |
|---|---|
| $y = \left( \dfrac{y}{x} \right)$ | $x$ |
| $y = \left( \dfrac{1}{3} \right)$ | $x$ |
| a. | 3 |
| b. | 15 |
| c. | 30 |
| d. | 1 |
| e. | 0 |

*Teaching note:* The *x* values are given. Students figure out corresponding *y* values.

## EQUATIONS FOR LINES THAT GO THROUGH ZERO (Lessons 55–65)

The work with function tables prepares students for using the general equation for straight lines that go through zero on the coordinate system. The exercise in lesson 55 relates the equation to the coordinate system. Here's part of the introduction:

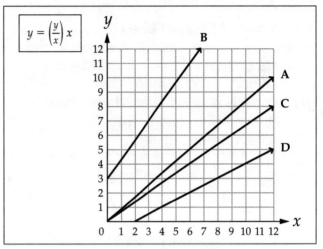

b. The last tables you worked with had an equation that told what Y equals. You can see that equation in the box: Y = Y over X times X.
- Everybody, say that equation. **(Signal.)** *Y = (Y/X) X.*
- This equation refers to all lines on the coordinate system that go through **zero.**
- Which point do the lines go through? **(Signal.)** *Zero.*
  Yes, the point for zero.

c. Touch zero on the first coordinate system. √
- Do all the lines shown go through zero? **(Signal.)** *No.*
- So not all of these lines are covered by the equation Y = (Y/X) X.

d. Touch line A. √
- Everybody, does it go through zero? **(Signal.)** *Yes.*
- So the equation Y = (Y/X) X tells about all the points on that line.

e. Touch line B. √
- Does it go through zero? (Signal.) *No.*
- So does the equation Y = (Y/X) X tell about all the points on that line? (Signal.) *No.*

f. Raise your hand when you know which other line is covered by the equation Y = (Y/X) X. √
- Which line? (Signal.) *Line D.*
- Yes, line D goes through zero. So all the points on that line are covered by the equation Y = (Y/X) X.

g. The next box gives information about a line that goes through zero.
- (Teacher reference:)

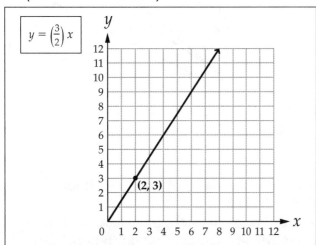

- One point on the line is X = 2, Y = 3.
- You can see the equation with the fraction 3/2:
  Y = 3 halves times X.
- Everybody, say that equation. (Signal.) *Y = (3/2) X.*
- You can see the line for that equation. It goes through the point X = 2 and Y = 3, and it also goes through zero.
- The equation Y = 3 halves X gives the rule for all the points on the line.

Here's part of the exercise in lesson 55 that addresses the relationship between the description of the point and the fraction:

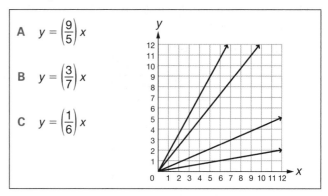

- Each of these equations tells about 1 of the lines. You're going to use the equations A through D to plot a point on each line.

b. Everybody, read equation A. (Signal.) *Y = (9/5) X.*
- Say the fraction. (Signal.) *9/5.*
- Which number is the X value? (Signal.) *5.*
- Which is the Y value? (Signal.) *9.*
- (Repeat step b until firm.)

c. Everybody, read equation B. (Signal.) *Y = (3/7) X.*
- Say the fraction. (Signal.) *3/7.*
- Which number is the X value? (Signal.) *7.*
- Which is the Y value? (Signal.) *3.*
- (Repeat step c until firm.)

d. Everybody, read equation C. (Signal.) *Y = (1/6) X.*
- Say the fraction. (Signal.) *1/6.*
- Which number is the X value? (Signal.) *6.*
- Which is the Y value? (Signal.) *1.*
- (Repeat step d until firm.)

In lessons 57 and 58, students complete a function table and plot the points. This table presents $y$ values on the left, so it matches the order of values in the equation.

To figure out the missing $x$ or $y$ values, students use the equation

$$y = \left(\frac{y}{x}\right) x.$$

In lessons 59–65, students work various problems involving slope.

Here's part of the introduction of slope from lesson 59:

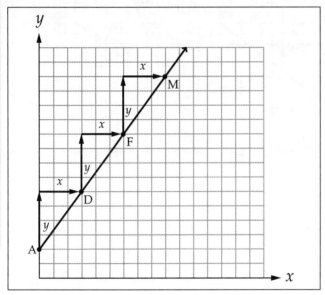

b. You're going to learn about the slope of a line.
• What are you going to learn about? (Signal.) *The slope of a line.*
c. I'll read. Follow along.
• The slope of a line is the fraction Y over X.
• What's the fraction that shows the slope of a line? (Signal.) *Y/X.*
• The slope tells how many units Y changes for the corresponding change in X.
• If the fraction for the slope is 4 thirds, Y increases 4 as X increases 3.
• That means the line goes up 4 units as it goes across 3 units. The diagram shows a slope of 4/3.

d. Touch point A on the Y axis. √
• The arrow for Y goes up. Touch that arrow. √
• That shows how many units Y changes. Count how many units that arrow goes up. √
• Everybody, how many units? (Signal.) *4.* Yes, Y increases 4.
e. Touch the corresponding X arrow. √
• That shows how many units X changes.
• Count how many units that arrow goes across. √
• Everybody, how many units? (Signal.) *3.* Yes, X increases 3.
• So Y increases 4, and X increases 3. That's a slope of 4 thirds.
f. Touch the point where the first X arrow touches the line. √
• Everybody, what's the letter of that point? (Signal.) *D.*
g. You can see the other pairs of arrows follow the same pattern. Touch the next point on the line where an X arrow touches the line. √
• What's the letter of that point? (Signal.) *F.*
h. Touch the last point where an X arrow touches the line. √
• What's the letter of that point? (Signal.) *M.*
i. So D, F, and M are points on the line.
• Each pair of arrows shows that the same thing is repeated. You start at the line. You go up 4 for Y. Then you go across 3 for X. And you're back at the line.

*Teaching note:* The coordinate system is shown without numbered axes so that students do not try to relate the slope to numbers on the axes. The convention of unnumbered axes continues in lessons that immediately follow, after which students return to the standard coordinate system.

In lesson 61, students figure out the slope of lines and write the equations.

In the following lessons, students convert equations such as

$$\frac{2}{3}y = 6x$$

into the form

$$y = \left(\frac{y}{x}\right)x.$$

Students solve for $y$, simplify, and write an equation for the slope. Here's the work for the example above:

$$\left(\frac{3}{2}\right)\frac{2}{3}y = 6x\left(\frac{3}{2}\right)$$

$$\boxed{y = 9x}$$

$$\boxed{\text{slope} = 9}$$

In lesson 64, the exercise introduces the standard equation for lines that go through zero:

$$y = mx.$$

Students solve equations for y, then write a simple equation for the slope:

$$m = \blacksquare.$$

> **Teaching note:** Remind students that $m$ is $y$ over $x$, so solving for $m$ is solving for $y$ over $x$. If the resulting fraction is $\frac{1}{2}$, the $y$ value is 1 and the $x$ value is 2. If $m = 6$, the $y$ value is 6 and the $x$ value is 1, not zero.

## THE EQUATION FOR LINES THAT DO NOT GO THROUGH ZERO (Lessons 66–81)

In lesson 66, students learn the general equation

$$y = mx + b.$$

The new part of the equation, $+ b$, is the intercept of the line at the $y$ axis.

The introductory exercise presents a coordinate system with four lines.

$$y = \frac{7}{3}x + b \qquad y = \frac{2}{5}x + b$$
$$y = \frac{3}{4}x + b \qquad y = \frac{3}{7}x + b$$

Students match each equation with the corresponding line (based on the slope), then complete the equation by writing the correct $b$ term. In lessons 67 and 70, students write complete equations for lines—showing both the slope and the intercept.

In lesson 74, negative slopes are introduced. Students learn the rule that when the slope is negative, the $y$ value decreases as the $x$ value increases. Students write fractions to show that the change in $y$ is negative, for example, $\frac{-2}{+3}$. The resulting fraction is negative: $-\frac{2}{3}$.

For any pair of points on a line, students follow the procedure of starting with the point to the left and testing whether the change in y is positive or negative. If it's negative, the line has a negative slope.

In lessons 76 and 77, students write the equations for a full range of lines—those with either a positive or a negative slope and those with either a positive or a negative intercept.

In lessons 79–81, students plot lines based on equations. The procedure students follow is 1) plot a point where the line crosses the y axis, 2) plot another point based on the slope, and 3) draw a line to connect the points.

Here's part of the exercise from lesson 79:

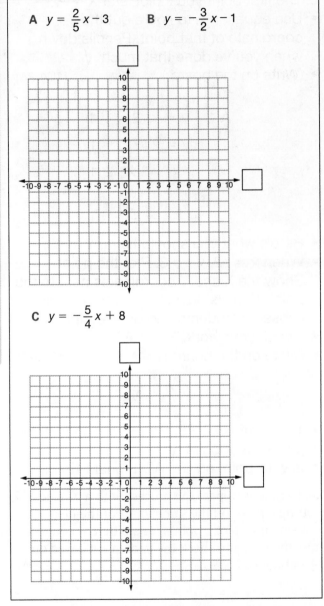

A $y = \frac{2}{5}x - 3$     B $y = -\frac{3}{2}x - 1$

C $y = -\frac{5}{4}x + 8$

- Label the X and Y axes. √
- Remember, first you'll plot the point on the Y axis.

b. Equation A: Y = 2 fifths X − 3.
- Where does the line cross the Y axis? (Signal.) − 3.
c. Equation B: Y = − 3 halves X − 1.
- Where does the line cross the Y axis? (Signal.) − 1.
d. Equation C: Y = − 5 fourths X + 8.
- Where does the line cross the Y axis? (Signal.) + 8.
- (Repeat steps b–d until firm.)

e. Go back to equation A.
- Once more: Where does the line cross the Y axis? (Signal.) − 3.
- After you plot the point on the Y axis, you make a second point based on the slope. What's the slope? (Signal.) 2/5.
- Do you move up or down for Y? (Signal.) Up.
- Do you move left or right for X? (Signal.) Right.
f. Equation B. What's the slope? (Signal.) − 3/2.
- Do you move up or down for Y? (Signal.) Down.
- Do you move left or right for X? (Signal.) Right.
g. Equation C. What's the slope? (Signal.) − 5/4.
- Do you move up or down for Y? (Signal.) Down.
- Do you move left or right for X? (Signal.) Right.
- (Repeat steps e–g until firm.)
h. Make the line for equation A. Remember, plot the 2 points. Then draw the line. Label it **A**. Pencils down when you're finished.
  (Observe students and give feedback.)
- Check your work.
- You should have 1 point at − 3 on the Y axis.
- The other point is up 2 places and to the right 5 places.

The final extension involving straight-line equations appears in lessons 103–106.

The problems give information about two points. Students solve for either *x* or *y*. Here's a problem from the second day of instruction in lesson 104.

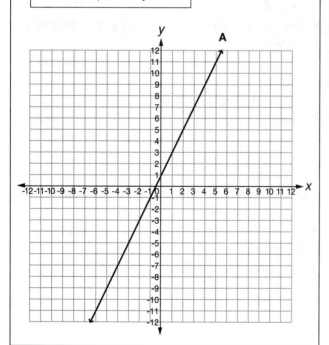

**Line A:**   $y = 2x + 1$

◆ Plot the point for $x = -2$.

◆ Plot the point for $y = 7$.

- You're going to plot points and draw a line for an equation.
b. The equation for line A is Y = 2X + 1.
- You're going to plot 2 points on the line, then use the equation to check where the line crosses the Y axis.
c. The first point you'll plot is at X = −2.
- Use equation A to figure out the Y coordinate of that point. Pencils down when you've done that much. √
- (Write on the board.)                    [104:2A]

> **A**   $y = \quad 2x + 1$
>
> $y = 2(-2) + 1$
>
> $y = \quad -4 + 1$
>
> $\boxed{y = \quad -3}$

- Here's what you should have. Y = −3.
- When x = −2, y = −3. Plot the point. Show the X and Y coordinates next to the point. Pencils down when you're finished. **(Observe students and give feedback.)**
- Check your work.
- (Write on the board:)                    [104:2B]

> $(-2, -3)$

- The point is −2, −3. Raise your hand if you plotted that point. √
- Is your point on line A? **(Signal.)** *Yes.*

Next, students repeat the process with a *y* value of 7, plotting a point on the same line. For the next problem, they plot 2 points for a different equation, draw the line, and check the *y* intercept based on the equation.

## GRAPHING RATE (Lessons 102–106)

In lessons 102–106, students work with rates represented as straight lines on the coordinate system. In lesson 102, they learn the conventions.

Here's part of the introduction:

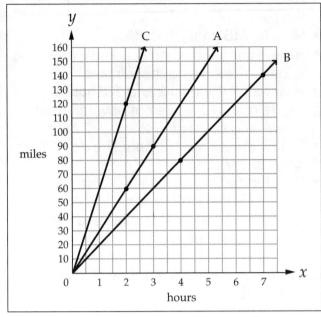

b. You can show rates on the coordinate system.
- The graph shows the rate **miles per hour.**
- The Y axis shows **miles.**
  Each unit on the Y axis shows 10 miles.
- The X axis shows hours.
  Each unit on the X axis shows 1 hour.
c. The lines show the rates of 3 cars.
- The rate is a straight line.
  The faster the rate, the steeper the slope.
d. Touch the line for the car with the fastest rate. √
- Everybody, which line are you touching? (Signal.) *C.*
  Yes, car C has gone the greatest distance after 1 hour.
- Raise your hand when you know how far car C has gone after 1 hour.
- Everybody, how far has it gone? (Signal.) *60 miles.*
e. Touch the line for the car with the slowest rate. √
- Which line are you touching? (Signal.) *B.*

- Raise your hand when you know how far car B has gone after 1 hour. √
- Everybody, how far has it gone? (Signal.) *20 miles.*
f. Touch the first point on line A. √
- The Y value for that point shows how far the car has gone. The X value shows how much time has gone by.
- Look at the Y axis. How far has the car gone? (Signal.) *60 miles.*
- Look at the X axis. How long did it take the car to go 60 miles? (Signal.) *2 hours.*
g. Touch the next point on line A. √
- Look at the Y axis. How far has the car gone? (Signal.) *90 miles.*
- Look at the X axis. How long did it take the car to go 90 miles? (Signal.) *3 hours.*
h. Line B shows the rate of another car.
- You can use the graph to figure out the rate of that car.
- Touch **1 hour** on the X axis and go up the line. There's no point shown there. How many miles did the car travel in that hour? (Signal.) *20 miles.*
  Yes, 20 miles. So the rate is 20 miles per hour.
i. Touch the first point on line B. √
- Look at the Y axis. How far has the car gone? (Signal.) *80 miles.*
- Look at the X axis. How much time did it take the car to go 80 miles? (Signal.) *4 hours.*
- Touch the next point on line B. √
- How far has the car gone? (Signal.) *140 miles.*
- How much time did it take? (Signal.) *7 hours.*
j. Touch line C. √
- Look at **1 hour** and figure out the rate of the car in miles per hour.
- What's the rate of the car? (Signal.) *60 miles per hour.*
- Touch the point on line C. √
- How far has the car gone? (Signal.) *120 miles.*
- How much time did it take? (Signal.) *2 hours.*

**Teaching note:** Following the part of the exercise shown above, students then answer questions about rate and questions about distance or time for lines representing the rate of growth for various trees.

In later lessons, students compare rates of different objects and answer questions that refer to the difference in Y values or the difference in X values.

Here's part of the exercise from lesson 103.

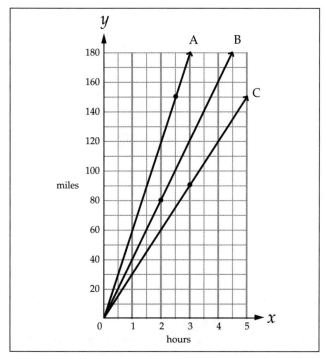

- You can use the graph to answer questions about each car. You can also answer questions that compare the cars.
c. Line A shows the rate for car A.
- Raise your hand when you know the rate of that car in miles per hour. √
- What's the rate of car A? (Signal.) *60 miles per hour.*
d. Raise your hand when you know the rate of car B. √
- What's the rate of car B? (Signal.) *40 miles per hour.*

e. Raise your hand when you know the rate of car C. √
- What's the rate of car C? (Signal.) *30 miles per hour.*
f. Touch the point on the line for car A. √
- Raise your hand when you know how far the car has gone. √
- How far? (Signal.) *150 miles.*
- Raise your hand when you know how long it took the car to go that distance. √
- How long? (Signal.) *2 and 1/2 hours.*
g. You're going to figure out how far apart cars A and B are after 2 and 1/2 hours.
- Follow the 2-and-1/2-hour line down from car A to car B. √
- How far has car B gone in 2 and 1/2 hours? (Signal.) *100 miles.*
  So car A has gone 150 miles and car B has gone 100 miles.
- You can do a subtraction problem to figure out how far apart they are: 150 – 100.
- What's the answer? (Signal.) *50.*
- Everybody, how far apart are the cars after 2 and 1/2 hours? (Signal.) *50 miles.*

In later lessons, students work various comparison problems involving rate of growth, rate of consumption, or rate of production. Students refer to the graph to answer questions about the difference, the rate, or the amount.

**COORDINATE SYSTEM RANGE (Lessons 114–118)**
In the final sequence involving the coordinate system, students identify the section of the coordinate system that satisfies conditions of X and Y. They plot points that would lie in that range. They identify points that would be outside the range or inside the range.

Here's part of the introduction from lesson 114:

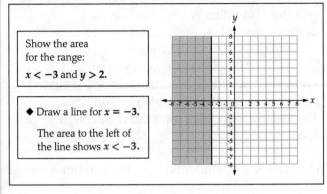

Show the area for the range:

$x < -3$ and $y > 2$.

◆ Draw a line for $x = -3$.

The area to the left of the line shows $x < -3$.

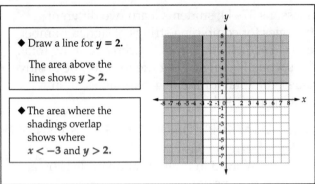

◆ Draw a line for $y = 2$.

The area above the line shows $y > 2$.

◆ The area where the shadings overlap shows where $x < -3$ and $y > 2$.

- Some problems ask about an area on the coordinate system.
- b. Here's a problem: Show the area for the range $Y > 2$ and $X < -3$.
- c. Here's how you work the problem.
- $X < -3$, so you draw a line for $X = -3$.
- Touch the line for $X = -3$. √
- The area to the **left** of the line shows $X < -3$.
- d. Now you do the same thing for Y.
- The problem tells that $Y > 2$. So you draw a line for $Y = 2$.
- Touch that line on the second coordinate system. √
- The area **above** the line shows $Y > 2$.
- e. The area where the shadings overlap shows where $X < -3$ **and** $Y > 2$.
- Touch that area. √
- So any point within that area could be $Y > 2$ and $X < -3$.
- f. Remember:
  - Draw a line for X, and draw a line for Y.
  - Then shade the areas that that the problem refers to.

Following the introduction, students shade areas for a range of X and Y and plot points within the range.

Exercises in later lessons present a coordinate system with various points. Students identify which points meet specific conditions. Here's the problem set from lesson 118:

a. Which point meets the conditions

$x < -4, y < -5$?

b. Which point meets the conditions

$x > 7, y > 0$?

c. Which point meets the conditions

$x < -1, y > 6$?

d. Which point meets the conditions

$x > 0, y < 1$?

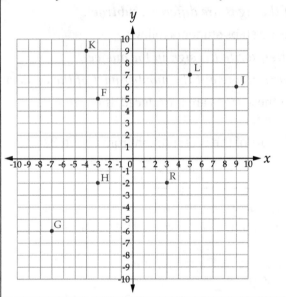

# Signed-Number Operations (Lessons 30–69)

Students learn rules for combining signed numbers, for multiplying and dividing signed numbers, and for following an order of operations. Students apply the rules to a variety of algebraic problems.

## COMBINING SIGNED NUMBERS (Lessons 30–44)

Signed numbers are first shown on a number line. Students learn that positive numbers are more than zero and negative numbers are less than zero. A plus value, like $+ 3$, tells both the direction from zero and the number of places:

3 places in the positive direction (to the right);

$- 7$ directs going seven places in the negative direction (to the left).

Students solve combining problems using a number line. They start at zero and perform two operations, one for each signed number. To work the problem + 3 − 7, they start at zero, go three places in the positive direction, and go 7 places in the negative direction.

This practice with the number line gives students the perspective needed to understand the rules for combining signed numbers. Students learn rules for the sign and the number part of the answer. The sign is the sign of the value farther from zero. These rules tell how to figure out the number part of the answer:

> *If the signs are the same, add.*
>
> *If the signs are different, subtract.*

These are steps students follow:

> *First, write the sign in the answer.*
>
> *Next, figure out the number part of the answer.*

When these steps are introduced, they are structured.

Here's part of the presentation from lesson 32:

| a. + 15 − 19 = |
|---|

f. You'll work each problem. First you'll put the sign in the answer. Then you'll figure out the number.

g. Problem A: Plus 15 minus 19.
- Write the sign in the answer. √
- Everybody, what sign? (Signal.) *Minus.*
- Yes, minus 19 is farther from zero. So the sign in the answer is minus.
- Now figure out the number part of the answer. Remember, see if you're going in the same direction for both operations or in the opposite directions. Pencils down when you're finished.
  **(Observe students and give feedback.)**
- Check your work.
- Did you add or subtract to find the number part of the answer? (Signal.) *Subtract.*
- You worked the problem 19 − 15.
- What's the number part of the answer? (Signal.) *4.*
- What does + 15 − 19 equal? (Signal.) *− 4.*
- Everybody, read the whole equation. (Signal.) *+ 15 − 19 = − 4.*

In lessons 32–36, students apply the combining rules to decimals and fraction values.

In lessons 37–44, students learn two different strategies for combining strings of signed number values. The first strategy is to keep a running total. The second is to figure out the total for the positive values, the total for the negative values, then combine the totals.

**MULTIPLICATION/DIVISION RULES (Lessons 48–57)**
Students learn the basic rules that if the sign of the numbers involved in multiplication or division are the same, the answer has a positive value; if the signs are different, the answer has a negative value.

In lesson 48, students apply the rules to multiplication problems. Here's part of the exercise that follows the introduction:

a. − 5 (− 2) =

b. + 10 (− 8) =

- These are signed number problems that multiply.
- b. Everybody, read problem A. (Signal.) *− 5 (− 2).*
- Are the signs the same or different? (Signal.) *Same.*
- What's the sign in the answer? (Signal.) *Plus.*
- The numbers are 5 and 2. What's 5 times 2? (Signal.) *10.*
- What's the whole answer? (Signal.) *+ 10.* Yes, − 5 (− 2) = + 10.
- Say that fact. (Signal.) *− 5 (− 2) = + 10.*

c. Read problem B. (Signal.) *+ 10 (− 8).*
- Are the signs the same or different? (Signal.) *Different.*
- What's the sign in the answer? (Signal.) *Minus.*
- The numbers are 10 and 8.
  What's 10 times 8? (Signal.) *80.*
- So what's the whole answer? (Signal.) *− 80.*
  Yes, *+ 10 (− 8) = − 80.*
- Say that fact. (Signal.) *+ 10 (− 8) = − 80.*

Students orally answer questions about three more examples, then work all problems.

> *Teaching note:* Expect some students to confuse the rules for combining and multiplying. Both sets of rules refer to signs that are the same and signs that are different. For combining, if the signs are the same, add. For multiplication, if the signs are the same, the answer is positive. The crux of the problem students may have is that although they add when combining values that have the same sign, the signs may both be negative and the answer negative. The simplest procedure for minimizing confusion is to remind students of the rules before they work sets of problems.

In lessons 54 and 55, students learn about division. The problems are written as fractions. Students are shown that the signed value for the numerator and for the denominator work just like signs for multiplication. If the signs are the same, the value is positive; if the signs are different, the value is negative.

In lesson 56, students work a mix of multiplication and division problems. Here's the problem set for lesson 56:

a. $\dfrac{-8}{+r} =$ ▮　　e. $-\dfrac{1}{2}\left(-\dfrac{5}{2}\right) =$ ▮

b. $+\dfrac{4}{3}\left(-\dfrac{2}{5}\right) =$ ▮　　f. $\dfrac{+10}{-1} =$ ▮

c. $\dfrac{-16}{-8} =$ ▮　　g. $\dfrac{-3}{-6} =$ ▮

d. $+ 5 (+ 6) =$ ▮　　h. $-\dfrac{6}{8}(+ 2) =$ ▮

## MORE THAN ONE OPERATION (Lesons 58–73)

In lessons 58–60, students work problems that have parentheses and that involve combining. They learn the order of operations: **First remove parentheses; then combine.**

In lessons 72 and 73, students work problems in which three numbers are multiplied. Students first figure out the signed product for two values, then multiply that value by the third value. Here's a problem from lesson 73:

c.　$− 4 − 2 (− 3) (+ 2) − 1 =$　▮
　　$− 4\qquad + 12\qquad − 1 =$　$\boxed{+ 7}$

Note that students copy the values that are not multiplied and show the products below the values that are multiplied.

## DISTRIBUTION WITH SIGNED-NUMBER OPERATIONS (Lessons 60–69)

In lesson 60, students learn to distribute in expressions like

$$− 5 (3r − 2 − 6b).$$

The procedure they follow is to multiply each term inside the parentheses by − 5.

Here's part of the presentation from lesson 60:

| **b.** $− 5 (3r − 2 − 6b)$ |
|---|

j. Problem B: − 5 times the quantity, 3R − 2 − 6B.
  You'll say the distribution problem for each term.
- Say the problem for the first term. (Signal.) *− 5 (3R).*
- What's the sign in the answer? (Signal.) *Minus.*
- What's the whole answer? (Signal.) *− 15R.*
- Say the problem for the next term. (Signal.) *− 5 (− 2).*
- What's the sign in the answer? (Signal.) *Plus.*
- What's the whole answer? (Signal.) *+ 10.*
- Say the distribution for the last term. (Signal.) *− 5 (− 6B).*
- What's the sign in the answer? (Signal.) *Plus.*
- What's the whole answer? (Signal.) *+ 30B.*
- (Repeat step j until firm.)

k. Copy the problem. Below, write the terms you get when you distribute. Pencils down when you're finished.
(Observe students and give feedback.)
- (Write on the board:)                                    [60:1B]

> **b.**    $- 5 (3r - 2 - 6b)$
>           $- 15r + 10 + 30b$

- Here's what you should have: $- 15R + 10 + 30B$.

*Teaching note:* The goal is to provide enough repetition for students to become relatively automatic in saying the problems they will work when distributing. Hold students to a high criterion, and repeat step j until students are firm.

In lesson 62, students follow the same procedure for problems that show the distributed value on the right side of the parentheses.

Here's part of the exercise from lesson 61:

> **b.**  $(- 2 + 4g + 6m) \dfrac{1}{2}$

c. Read expression B. (Signal.) *The quantity minus 2 + 4G + 6M, times 1/2.*
- What value is being distributed? (Signal.) *1/2.*
- Say the problem for the first term. (Signal.) *– 2 (1/2).*
- Say the problem for the next term. (Signal.) *+ 4G (1/2).*
- Say the problem for the next term. (Signal.) *+ 6M (1/2).*
- Copy expression B. Below, write the terms you get when you distribute 1/2. Show the fractions as whole numbers.
- (Write on the board:)                                    [61:1B]

> **b.**    $(- 2 + 4g + 6m) \dfrac{1}{2}$
>           $- 1 + 2g + 3m$

- Here's what you should have: $- 1 + 2G + 3M$.

*Teaching note:* Students are expected to apply what they have learned about simplifying fractions to this work. Some students may have to write the values that are to be simplified. If students require this work, make sure they indicate both the numbers and the signs. Remind them that an answer without a sign is shown in the distributed expression as a positive value.

Starting in lesson 62, students work problems in which they combine like terms before they distribute. Here's a problem from lesson 62:

$$- 8 (b - 30 + 15 - 6b).$$

Students first rewrite the problem with like terms combined:

$$- 8 (- 5b - 15).$$

Then students remove the parentheses:

> $40b + 120$

In lesson 63, students work problems that have a term that is not involved in the distribution:

$$5k - 10 (k + 4).$$

The steps students follow are first to work the part that is distributed and copy the part that is not involved in the distribution. Then students combine like terms.

> $3 (2m - 7) + 5$   ◄ Work the part that is distributed.
> $6m - 21 + 5$   ◄ Copy the part not involved in the distribution.
> $6m - 16$   ◄ Combine like terms.

Starting in lesson 64, students work with equations that involve distribution. First students distribute, then simplify and solve. Here's a problem with the solution steps shown:

**c.** $(4t + 1 - 5t)(-2) - 3 = 1$
$(-t + 1)(-2) - 3 = 1$
$2t - 2 - 3 = 1$
$2t - 5 = 1$
$\underline{+5 \quad +5}$
$\left(\dfrac{1}{2}\right) 2t = 6 \left(\dfrac{1}{2}\right)$
$\boxed{t = 3}$

*Teaching note:* Students are not to take shortcuts or skip steps. After the procedures have been established, you may provide more latitude, but if the students are not firm on the steps, they will make mistakes later.

Notice that two sets of parentheses show that the $(-2)$ is distributed. Students have learned that a signed value after parentheses must also have parentheses to show it is multiplied; otherwise, it is combined.

In lesson 66, students learn that a minus sign in front of a parentheses means that everything inside the parentheses is multiplied by minus one.

$10 = t - (-2t + 5)$
$10 = t + 2t - 5$

In lesson 67, students distribute a letter.

$r(-7 + 2k) + 2r$
$-7r + 2kr + 2r)$

In lessons 68 and 69, students work problems that require substitution and distribution (see pages 84–85 for information about how substitution is taught). The solution involves many steps and represents a significant achievement in applying algebraic operations.

Here's part of the exercise from lesson 68:

**a.** $21 = 6q + q(5 - 2r)$   $\boxed{r = 2}$

**b.** Problem A:
   $21 = 6Q + Q$ times the quantity $5 - 2R$.
   $R = 2$.
- Copy the problem and work it. Substitute for 1 of the letters. Simplify inside the parentheses and solve for the other letter. Pencils down when you're finished. **(Observe students and give feedback.)**
- (Write on the board:)   [68:5A]

**a.**   $21 = 6q + q(5 - 2r)$   $\boxed{r = 2}$
   $21 = 6q + q(5 - 2(2))$
   $21 = 6q + q(1)$
   $\left(\dfrac{1}{7}\right) 21 = 7q \left(\dfrac{1}{7}\right)$
   $\boxed{3 = q}$

- Here are the steps.
- R = 2. So inside the parentheses, you have $5 - 2$ times 2. That's $5 - 4$. So $21 = 6Q + 1Q$.
- What does Q equal? **(Signal.)** *3.*

*Teaching note:* The typical difficulties students encounter involve a small calculation that is wrong or a small step left out. If students tend to make many of these errors, but obviously understand what they are to do to solve problems, you may want to give them the answer to the problem and possibly the step before the answer.

This procedure provides students with a basis for checking their work and possibly finding the mistake. (Often, without this information, students don't have efficient procedures for finding mistakes).

## Decimal Division (Lessons 36–52)

Before this track begins, students have practiced reading fractions as division problems and working them—expressing the answer as a whole number or a mixed number. (See Fraction Equivalences, page 56.)

The decimal-division track expands on this foundation. The first activities involve fractions that have a decimal value only in the denominator. Students make the denominator a whole number by moving the decimal point. They move the decimal point the same number of places in the numerator. Then they work the problem as a whole-number division problem.

$$\frac{6}{.2} = \frac{6.0}{.2} \qquad 2\overline{)60}$$

Next, students work division problems that have two decimal values. Students learn to add a zero after the decimal point in the dividend to work problems that have a decimal remainder.

Students practice rounding answers to the nearest hundredth. Finally, students work with fractions that have decimal values in both numerator and denominator. Students follow the same practice of "clearing" the decimal point. Then they simplify the fraction.

In lessons 36–38, students read and work fraction problems as division problems.

Here's part of the introduction from lesson 36:

$$\frac{56}{.08}$$

$$\frac{56}{.08} \times \frac{100}{100} = \frac{5600}{8} \qquad \boxed{700} \atop 8\overline{)5600} \qquad \text{So: } \frac{56}{.08} = 700$$

a. $\dfrac{15}{.05}$

b. The first fraction in the first box is 56 over .08. This fraction has a whole number in the numerator and a decimal number in the denominator.
- The denominator shows hundredths. So we can get rid of the decimal point by multiplying both the numerator and the denominator by 100.
- You can see that in the numerator we get 5600. In the denominator, we get 8. Now we can work the division problem for that fraction.
c. The next box shows 5600 divided by 8 = 700. So, 56 over .08 = 700.
d. We can change fractions the fast way by moving the decimal point the same number of places in the numerator and the denominator.
e. Listen: If the number in the denominator has **1 place after the decimal point,** you move each decimal point **1** place.

- If the number in the denominator has **2** places after the decimal point, how many places do you move each decimal point? (Signal.) *2 places.*
f. Look at fraction A. √
- What's the denominator? (Signal.) *.05.* Yes, 5 hundredths.
- How many decimal places are in 5 hundredths? (Signal.) *2.*
- So how many places do you move each decimal point? (Signal.) *2.*

In lessons 36 and 37, students practice writing equations that show the original value and the equivalent fraction with the decimal point removed in the denominator.

Here's the problem set from lesson 37:

a. $\dfrac{7}{.05} = \dfrac{700}{5}$    c. $\dfrac{5.8}{.2} = \dfrac{58}{2}$    e. $\dfrac{4.25}{.025} = \dfrac{4250}{25}$

b. $\dfrac{2.75}{2.5} = \dfrac{27.5}{25}$    d. $\dfrac{8}{.004} = \dfrac{8000}{4}$    f. $\dfrac{8.05}{.5} = \dfrac{80.5}{5}$

In lesson 38, students write each fraction as a division problem, move the decimal points, and work the division problem. All examples yield a whole-number answer.

In lessons 39 and 40, students work division problems that have two decimal values and that result in a decimal answer. Students move the decimal points to create a whole-number divisor. They next write the new decimal point in the answer and work the problem.

Here's part of the introduction:

a. $.5\overline{)3.55}$

b. Problem A: 3.55 divided by .5.
- Say the problem. (Signal.) *3.55 ÷ .5.*
- What are you dividing by? (Signal.) *.5.*
- You must divide by a whole number. So how many places do you move the decimal point? (Signal.) *1.* Yes, 1 place.
- (Write on the board:)                    [39:3A]

a.    $.5\overline{)3.5.5}$

- Here's problem A with the decimal points moved 1 place.

- You'll write the decimal point in the answer, right above the new decimal point.
- (Write to show:) [39:3B]

a. $5\overline{)3.5.5}$

- Copy the problem on the board and work it. Pencils down when you're finished. (Observe students and give feedback.)
- (Write to show:) [39:3C]

a. $\dfrac{7.1}{5\overline{)3.5.5}}$

- Here's what you should have. You divided 35.5 by 5. The answer is 7.1. That's the answer to the original problem.

*Teaching note:* You may have to remind students that they first move both decimal points the same number of places, just as they did with fractions. Then they write the new decimal point in the answer before they do any calculation.

In lessons 41–44, students add zeros to find decimal remainders. Students first determine if there is a remainder. If so, students figure out the decimal remainder. Students work problems that have decimal remainders to the tenths, hundredths, and thousandths. If there is no remainder after working the problem to the tenths, hundredths, or thousandths place, students stop there. If there is still a remainder after the thousandths place, students round to hundredths.

Here's part of the introduction from lesson 41:

$\dfrac{8.9}{5\overline{)4.4.7_{4}2}}$  ◄ There's a remainder of 2.

$\dfrac{8.94}{5\overline{)4.4.7_{4}0}}$  ◄ Write zero and divide again.

a. $.04\overline{).0380}$

- Some division problems that divide by a decimal value have a remainder. You won't show the remainder as a fraction. Instead, you'll show that remainder as a decimal value.
b. The first problem in the box shows 4.47 divided by .5.
- How many places do we move each decimal point to make 5 tenths a whole number? (Signal.) *1.*
- The new problem is 44.7 divided by 5.
- Now we work the problem: 5 goes into 44, 8 times. The remainder is 4. So you work the problem: 5 goes into 47 how many times?
- The answer is 9 with a remainder of 2.
- We don't write that remainder as a fraction, because we're working with decimal values.
- So we write zero and divide again. You can see the zero in the second division problem.
- The new problem is 5 goes into 20 how many times?
- What's the answer? (Signal.) *4.*
c. Remember, if you get a remainder after the decimal point, you make a zero and divide again.
d. Copy problem A. Move the decimal points so you're dividing by a whole number. Work the problem. You'll get a remainder after the last digit of the number you're dividing, so just make a zero and divide again. Pencils down when you're finished. (Observe students and give feedback.)
- (Write on the board:) [41:2A]

a. $\dfrac{.95}{.04\overline{).03.8_{2}0}}$

- Here's what you should have. You divide 3.8 by 4. You get a remainder of 2. So you add a zero and work the problem: 4 goes into 20 how many times? The answer is 5.
- What does 3.8 divided by 4 equal? (Signal.) *.95.*

In lesson 45, students work with mixed sets of decimal problems that call for addition, subtraction, multiplication, and division.

Students learn to simplify fractions with decimal values in lessons 51 and 52. Students first move the decimal point the same number of places needed to make both numerator and denominator whole numbers, then simplify.

Here's part of the introduction:

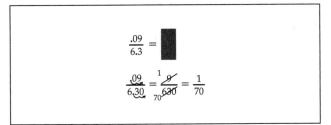

d. The fraction in the next box is **.09** over 6.3.
- How many places do you move the decimal points to make a whole number in the numerator and in the denominator? (Signal.) *2.*
- Yes, if you move the decimal points 1 place, the numerator will still have a decimal point.
- Say the fraction you get when both decimal points are moved 2 places. (Signal.) *9/630.*
- Now you can simplify. You can see that the fraction equals 1/70.

## Geometry (Lessons 45–118)

The geometry track focuses on circles, angles, similar triangles, Pythagorean theorem, surface area, and volume.

### CIRCLES (Lessons 45–60)

Students work with two values for *pi:* the rounded value, 3.14, and the calculator value.

**Students will need calculators with a *pi* key, π, starting in lesson 48.**

This track assumes that students know nothing about the nomenclature or equations for working with circles. Therefore, the track has exercises on circumference, diameter, and the equation $C = 3.14d$ for six lessons, 45–50.

In lesson 48, students are introduced to the symbol π and work problems using a calculator. They round the answer to hundredths.

**Note:** Students have learned pi as 3.14 and learned the equation for the circumference of a circle.

Here's part of the introduction from lesson 48:

$$\pi = 3.141592654$$

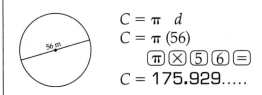

b. You can see the symbol for **pi** in the box. It says: Pi = 3.141592654.
- That's a number for **pi** that is not rounded to hundredths. Some calculators use that number.
- You're going to use a calculator with a **pi** key to work these problems.
c. The problem in the box shows a circle with a diameter of 56 meters.
- The equation with letters is C = π D.
- Below is the equation with a number for the diameter.
- You can see the keys you press on a calculator to work the problem.
- You enter **pi times 56** and **equals.**
- Work the problem on your calculator. (Observe and assist students. Give feedback.)
  Your answer is 175.929 and more digits.
- Write that number rounded to hundredths and box it. √
- (Write on the board:)                                          [48:4A]

175.93

- Here's what you should have.
  [Answer is 175.84 if 3.14 is used.]

In lesson 49, students find the diameter of circles. For this activity, they will probably need a calculator.

Here's part of the introduction:

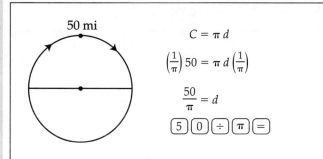

50 mi

$C = \pi d$

$\left(\frac{1}{\pi}\right) 50 = \pi d \left(\frac{1}{\pi}\right)$

$\frac{50}{\pi} = d$

⑤ ⓪ ÷ π =

b. Everybody, what's the name for 3.14? (Signal.) *Pi.*
• Say the equation for finding the circumference of a circle. (Signal.) $C = \pi D.$
• (Repeat step b until firm.) Yes, C = π D.
c. You can use the same equation for finding the diameter if you know the circumference.
d. Look at the problem in the box.
• The number and unit name for the circumference is shown. What's the circumference? (Signal.) *50 miles.*
• You can see the equation with letters: C = π D.
• Say that equation. (Signal.) $C = \pi D.$
• Below, you can see the equation with a number for C. To work the problem, you multiply both sides by 1 over pi.
• Everybody, what do you multiply both sides by? (Signal.) *1 over π.* That's 1 divided by pi.

• So on the side with 50, you have 50 divided by pi. On the other side, you have D.
• At the bottom of the box, you can see the keys you press to work the problem. Use your calculator. Use the π key. Round the answer to hundredths. Remember the unit name. Box your answer. Pencils down when you've figured out the diameter. **(Observe students and give feedback.)**
• (Write on the board:)                    [49:4A]

15.92 mi

• Here's what you should have. The diameter is 15.92 miles.

Radius is introduced in lesson 53. In lesson 54, the equation for area is introduced. The initial equation is

$$A = \pi r^2.$$

Here's part of the introduction:

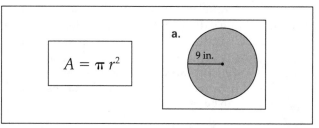

$A = \pi r^2$

a.

9 in.

b. You're going to figure out the area of circles.
• The box gives the equation for the area of a circle: A = pi R to the second.
• What's the equation for the area of a circle? (Signal.) $A = \pi r^2.$
• Remember, R to the second is **R times R.**

c. Here's the equation for a circle with a radius of 8 inches: A equals π times 8 times 8.
- Say the equation for the area of a circle with a radius of 8 inches. (Signal.) *A = π × 8 × 8.*

d. Say the equation for the area of a circle with a radius of 17. (Signal.) *A = π × 17 × 17.*

e. Say the equation for the area of a circle with a radius of 60. (Signal.) *A = π × 60 × 60.*

- (Repeat steps c–e until firm.)

f. Circle A. The number for the radius is shown. What's the number? (Signal.) *9.*

- Copy the equation in the box. Below, write the equation with a number for radius times radius. Stop when you've written both equations.
(Observe students and give feedback.)

> ***Teaching note:*** Students read R² as R to the second. This convention is used so that students express powers in a uniform way (as they would for R to the eighth). There is no particular problem with referring to "R squared" so long as the students understand that it is R times R. If you introduce an alternative way of reading, you may have sloppy student responses to items that have the exponents 2 and 3.

Starting in lesson 58, students find the circumference and the area, given the diameter or the radius. Here's the problem set from lesson 58:

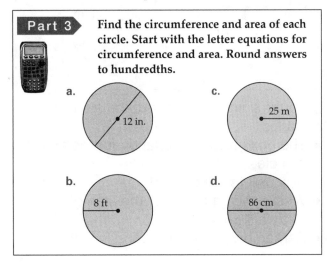

**Part 3** Find the circumference and area of each circle. Start with the letter equations for circumference and area. Round answers to hundredths.

a. 12 in.
b. 8 ft
c. 25 m
d. 86 cm

**ANGLES (Lessons 71–76)**

In lessons 71 and 72, students learn what angles are and learn facts about angles.

- The angle for the corner of a rectangle is 90 degrees.
- The angle for a straight line is 180 degrees.
- The angle for a circle is 360 degrees.

Students also learn that there are four 90-degree angles in a circle.

In lesson 72, students learn facts about triangles. Students also work problems in which two angles of a triangle are given; students find the number of degrees in the third angle.

Here's a part of the exercise:

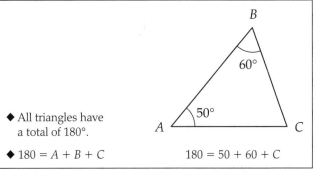

- ◆ All triangles have a total of 180°.
- ◆ 180 = A + B + C          180 = 50 + 60 + C

b. The rule in the box says: All triangles have a total of 180 degrees.
- Say that rule. (Signal.) *All triangles have a total of 180°.*

c. You can see the equation with letters for the angles: 180 = A + B + C.
- That means that angle A plus angle B plus angle C add up to 180 degrees for any triangle.
- If you know 2 angles of a triangle, you can figure out the number of degrees in the third angle.

d. You can see a triangle with angles A, B, and C.
- The triangle shows numbers for 2 of the angles. Which angles? (Signal.) *A and B.*
- How many degrees is angle A? (Signal.) *50 degrees.*
- How many degrees is angle B? (Signal.) *60 degrees.*
- So you can figure out angle C.
- You can see the equation below the triangle: 180 = 50 + 60 + C.

- Copy the equation and solve for C. Don't write the little circle for degrees. Pencils down when you're finished.
  (Observe students and give feedback.)
- (Write on the board:)                                    [72:3A]

$$180 = 50 + 60 + C$$
$$180 = 110 \quad\quad + C$$
$$-110 \quad -110$$
$$\overline{70 = \quad\quad\quad\quad C}$$

- Here's what you should have.
- Everybody, what does C equal? (Signal.) *70.*
- So angle C is 70°.

In lesson 73, students are introduced to the equation

$$m \angle D = \blacksquare °.$$

They read the equation as "The measure of angle D is so many degrees."

Here's part of the exercise that follows the introduction:

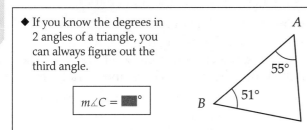

◆ If you know the degrees in 2 angles of a triangle, you can always figure out the third angle.

$$m \angle C = \blacksquare °$$

55°
51°
A
B
C

b. If you know the degrees in 2 angles of a triangle, you can always figure out the third angle.
- How many degrees are in any triangle? (Signal.) *180.*
  Yes, the sum of the 3 angles is 180 degrees.
- Let's say that the 2 angles you know add up to 100 degrees.
- Raise your hand when you know the third angle. √
- Everybody, what's the third angle? (Signal.) *80°.*
c. You can see a triangle with letters for the 3 angles.

- The angles are A, B, and C. The diagram shows the number of degrees for 2 angles. What are the letters for those angles? (Signal.) *A and B.*
- Which angle do you have to figure out? (Signal.) *C.*
- Work the problem. Figure out how many degrees are in angle C. Remember, start with the equation that starts with 180.
- (Write on the board:)                                    [73:3A]

$$180 = 55 + 51 + C$$
$$180 = 106 \quad + C$$
$$-106 \quad -106$$
$$\overline{74 = \quad\quad\quad\quad C}$$

- Here's what you should have.
- Everybody, how many degrees is angle C? (Signal.) *74.*
- (Write to show:)                                         [73:3B]

$$180 = 55 + 51 + C$$
$$180 = 106 \quad + C$$
$$-106 \quad -106$$
$$\overline{74 = \quad\quad\quad\quad C}$$

$$\boxed{m \angle C = 74°}$$

d. Here's how you show the answer. The sign is an angle marker. The equation says: The measure of angle C equals 74°.
- Everybody, read the equation. (Signal.) *The measure of angle C equals 74°.*
- What does M stand for? (Signal.) *Measure.*
- What does the marker after M stand for? (Signal.) *Angle.*
- What is the symbol after 74? (Signal.) *Degrees.*
  Yes, a degree symbol.
- (Repeat step d until firm.)
e. Copy the equation and box it. √
f. Remember, from now on, that's the way you'll write equations to show the number of degrees in an angle. M angle equals so many degrees.

In lesson 75, students learn that a 90-degree angle is called a **right angle.** They also learn how to figure out the missing angle in a right triangle. They learn that a right triangle has only one right angle and that the other two angles add up to 90 degrees. Students also learn that they can find the missing angle in a right triangle by subtracting the known angle from 90 degrees.

If students don't catch on to the idea of why subtracting the known angle from 90 yields the missing angle, tell them to do it the long way, working the problem,

$$180 = 90 + 35 + F.$$

Point out that they get the same answer by subtracting 35 from 90.

## SIMILAR TRIANGLES (Lessons 85–101)

The introduction in lessons 85–87 presents the description that similar triangles are the same shape, which means they have the same three angles. If you know two angles in two triangles, you can determine whether they have the same third angle and therefore whether they are similar.

Here's part of the introduction:

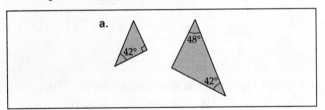

c. Item A. The angles given for the smaller triangle are 90° and 42°.
• The angles given for the larger triangle are 42° and 48°.
• One angle is the same for both triangles. So if you figure out the missing angle in **either** triangle, you'll know if they are similar.
• Raise your hand when you know the missing angle in the larger triangle. √
• What's the missing angle? (Signal.) *90°.* Yes, 90°.
• We know all 3 angles for the larger triangle. We know that the smaller triangle has 2 of them, 42° and 90°. So the missing angle in the smaller triangle **must** be 48°.

• Figure out the missing angle in the smaller triangle and see if it's 48°. √
• (Write on the board:)                    [85:4A]

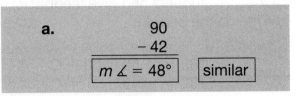

• Everybody, is the third angle 48°? (Signal.) *Yes.*
• So are the triangles similar? (Signal.) *Yes.*

Students practice with other pairs of triangles and determine whether each pair is similar.

> *Teaching note:* Students may use different strategies for finding the missing angles. The main point students are to learn is that if two angles in a pair of triangles are the same, the triangles are similar.

In lesson 88, students figure out the length of corresponding sides in similar triangles. Students construct a pair of equivalent fractions. The first shows the missing side in the numerator.

Following the introduction, students apply the conventions for determining the length of a corresponding side. In this exercise, students first set up all the problems, then go back and work them.

Here's the part of the exercise that directs them to set up the first problem:

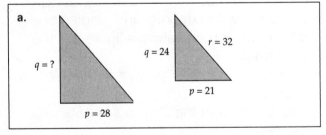

• Each pair of triangles is similar.
b. Problem A. We'll write an equation showing the pair of fractions you'll use to figure out the length of side Q in the larger triangle.
• Write a fraction for the corresponding Q sides. Remember, the side with the question mark goes in the numerator. √

- Everybody, what's the fraction for the corresponding Q sides? (Signal.) *Q over 24.*
- Now write an equal sign and complete the equation for the corresponding P sides. √
- (Write on the board:)                    [88:1A]

a.  $\dfrac{q}{24} = \dfrac{28}{21}$

- Here's what you should have: Q over 24 = 28/21.

Here's the work for the problem:

a. $(24)\ \dfrac{q}{24} = \dfrac{28}{21}\ (24)$

$$\boxed{q = 32}$$

**Teaching note:** The procedure of writing the unknown as the numerator of the first fraction reduces solution steps and is also consistent with what students are doing in both algebra word problems and problems that call for rate equations.

In lesson 89, students learn to describe sides of triangles by referring to the letters for the angles. Triangles with angles A, B, and C are shown to have sides AB, BC, and AC.

In lesson 92, the pairs of triangles are not oriented the same way, so students have to determine which sides of one triangle correspond to sides of the other. The routine they are to follow is to find the pair of longest sides first, then the pair of shortest sides.

In lesson 95, students work with nested triangles. These problems present information about parts of a corresponding side. Students add to find the length of a side for the larger triangle. Here's the first part of the exercise that follows the introduction:

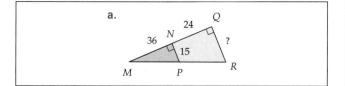

- Each item gives numbers for 1 pair of corresponding sides, but you may have to add to figure out the length of the longer corresponding side.
b. Item A. Find the question mark and write the fraction for the corresponding sides. √
- (Write on the board:)                    [95:4C]

a.  $\dfrac{QR}{15}$

- What's the fraction? (Signal.) *QR/15.*
- Touch the corresponding sides that have numbers. √
- What are those numbers? (Signal.) *36, 24.*
- The length of the shorter corresponding side with a number is 36.
- What's the length of that side? (Signal.) *36.*
- What do you have to add to get the length of the longer corresponding side? (Signal.) *36 + 24.*
- Raise your hand when you know the number for the longer corresponding side. √
- What number? (Signal.) *60.*
- Now complete the equation with a fraction for those corresponding sides. Then stop. √
- (Write to show:)                    [95:4D]

a.  $\dfrac{QR}{15} = \dfrac{60}{36}$

- Here's what you should have: QR/15 = 60/36.

A variation of nested triangles is introduced in lesson 97. For these, students subtract to find the length of the shorter corresponding side. Here's an example:

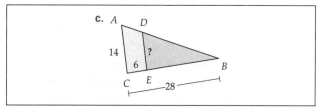

Word problem applications are introduced in lessons 100 and 101. The diagram is shown for these problems.

Here's part of the exercise from lesson 100:

a. The diagram shows 2 poles and their shadows. The length of shadow A is 10 feet. Pole B is 30 feet taller than pole A. Pole A is 15 feet tall. What is the length of shadow B?

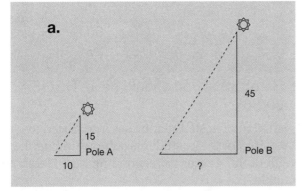

Pole A    Pole B

b. Problem A: The diagram shows 2 poles and their shadows. The triangles in the diagram are similar. The length of shadow A is 10 feet. Pole B is 30 feet taller than pole A. Pole A is 15 feet tall. What is the length of shadow B?

c. Put a question mark on the side of the diagram you have to figure out. Write the 3 numbers that the problem tells about. You'll have to add to figure out the height of pole B. Stop when you have 3 numbers and a question mark on the diagram. (Observe students and give feedback.)

• (Write on the board:)    [100:1A]

a.
45
15
Pole A    Pole B
10    ?

• Here's what you should have.
d. Write an equation with 3 numbers and a question mark for the side you'll figure out. Then figure out the length of shadow B. Pencils down when you're finished. (Observe students and give feedback.)

• (Write on the board:)    [100:1B]

$$\text{a.} \quad (10)\frac{?}{10} = \frac{45}{15}(10)$$
$$? = 30$$
$$\boxed{30 \text{ feet}}$$

• Here's what you should have.
• What's the length of shadow B? (Signal.) *30 feet.* Yes, 30 feet.

*Teaching note:* In a traditional sequence, this kind of problem would be a challenge for both teacher and students because a lot of teaching would have to take place before students would understand how to go about solving the problem. In the sequence above, however, the problem is a simple extension of what students already know.

On the final structured day of work with similar triangles, students work problems presented in the textbook. Problems show part of the diagram needed to solve the problem. Students copy that part and complete it using information from the problem. Here's a problem and solution from lesson 101:

a. Two cyclists, Bill and Jake, start out the same distance from a campsite. In the morning they ride 48 miles and stop for lunch. They are then 16 miles apart, and both are 24 miles from the campsite. How far apart are the cyclists when they start out?

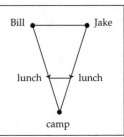

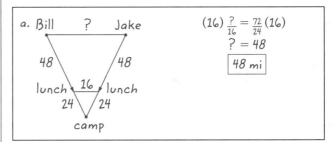

**PYTHAGOREAN THEOREM (Lessons 78–102)**

A preskill needed to work problems involving the Pythagorean theorem is square root, which is introduced in lesson 78. In lesson 86, students learn the equation $x^2 + y^2 = h^2$. In the following lessons, students solve problems for $x$, $y$, and $h$. Word problems that ask about $x$ or $h$ are introduced in lesson 97. In lesson 98, students discriminate between word problems that involve features of right triangles or word problems that describe straight lines (for instance, cars moving in the same direction or opposite directions).

In lesson 78, students find the square root. They learn the definition of square root and how to read and interpret problems that ask for the square root. For $\sqrt{36}$, students work this problem: What number multiplied by itself equals 36? For all the early items, the square root is a whole number.

In lesson 79, students work problems that require finding the square root and problems that require finding the squared value. Here's part of the introduction from lesson 79:

---
a. $\sqrt{\blacksquare} = 12$    c. $\sqrt{\blacksquare} = 5$    e. $\sqrt{\blacksquare} = .6$

b. $\sqrt{\blacksquare} = 13$    d. $\sqrt{\blacksquare} = 7$
---

b. Problem A asks a different kind of question: The square root of what number equals 12?
- What does problem A ask? (Signal.) *The square root of what number equals 12?*
- What does problem B ask? (Signal.) *The square root of what number equals 13?*

c. These problems are easy to work. For problem A, you just multiply 12 by itself. Copy the equation and complete it. Figure out what 12 times 12 equals, and write it under the square-root sign. Pencils down when you're finished.
(Observe students and give feedback.)
- (Write on the board:)                    [79:3A]

---
**a.**    $\sqrt{144} = 12$
---

- Here's what you should have: The square root of 144 equals 12.

d. Work problem B. Pencils down when you're finished. $\sqrt{}$
- (Write on the board:)                    [79:3B]

---
**b.**    $\sqrt{169} = 13$
---

- Here's what you should have: The square root of 169 equals 13.

e. You'll **read** the rest of the problems.
- Problem C. (Signal.) *The square root of what number equals 5?*
- Problem D. (Signal.) *The square root of what number equals 7?*
- Problem E. (Signal.) *The square root of what number equals .6?*

f. Copy and complete the rest of the equations in part 5. Pencils down when you're finished.
(Observe students and give feedback.)

g. Check your work.
- (Assign 3 students to show the work for equations C–E on the board. Students are to work simultaneously.)

**Key for board:**

c. $\sqrt{25} = 5$        e. $\sqrt{.36} = .6$

d. $\sqrt{49} = 7$

In lesson 82, students work from a table that shows square root whole-number values for 1–20.

---
| 1 | 2 | 3 | 4 | 5 | 6 | 7 |
|---|---|---|---|---|---|---|
| $\sqrt{1}$ | $\sqrt{4}$ | $\sqrt{9}$ | $\sqrt{16}$ | $\sqrt{25}$ | $\sqrt{36}$ | $\sqrt{49}$ |

| 8 | 9 | 10 | 11 | 12 | 13 | 14 |
|---|---|---|---|---|---|---|
| $\sqrt{64}$ | $\sqrt{81}$ | $\sqrt{100}$ | $\sqrt{121}$ | $\sqrt{144}$ | $\sqrt{169}$ | $\sqrt{196}$ |

| 15 | 16 | 17 | 18 | 19 | 20 |
|---|---|---|---|---|---|
| $\sqrt{225}$ | $\sqrt{256}$ | $\sqrt{289}$ | $\sqrt{324}$ | $\sqrt{361}$ | $\sqrt{400}$ |
---

Students work problems that do not involve whole-number answers and identify the two whole numbers closest to the square-root value.

Here's part of the introductory exercise:

f. Listen: The square root of 150 is between 2 square root numbers on the chart.
- Touch the 2 square roots 150 is between. √
- The square root of 150 is between the square root of 144 and the square root of 169.
- What's the square root of 144? (Signal.) *12.*
- What's the square root of 169? (Signal.) *13.*
- So the square root of 150 is between which whole numbers? (Signal.) *12 and 13.*
- (Repeat step f until firm.)

In the same lesson, students use their calculator to find square root values that are not whole numbers and round them to hundredths.

In lessons 83 and 84, some of the problems students work have a whole-number square root; others don't. Students identify the whole root number or indicate the two whole numbers closest to the square-root value.

The first work on the Pythagorean theorem begins in lesson 86. The exercise introduces information about the sides of a right triangle. Students learn that the hypotenuse is the longest side of the triangle—the side opposite the right angle.

The exercise next introduces the equation for right triangles:

$$x^2 + y^2 = h^2.$$

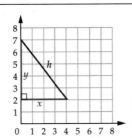

- The side **opposite the right angle** is the **hypotenuse.**
- The **hypotenuse** is always the **longest side** of the triangle.

$$x^2 + y^2 = h^2$$
$$4^2 + 5^2 = h^2$$
$$16 + 25 = h^2$$
$$\boxed{41 = h^2}$$

e. The box shows a rule for the length of the sides of a right triangle.
- X squared plus Y squared equals H squared.
- Say that rule. (Signal.) $X^2 + Y^2 = H^2.$
- (Repeat until firm.)

f. That does not mean that you can add up the lengths of X and Y to get the hypotenuse. It means that the length of X **squared** plus the length of Y **squared** gives you the length of H **squared.**
- Does X squared plus Y squared give you the length of H? (Signal.) *No.*
- What does it give you? (Signal.) *The length of H squared.*
- (Repeat until firm.)
- Yes, the length of H squared.

g. The coordinate system shows an example.
- Touch the side labeled X. √
- How many units is X? (Signal.) *4.*
- Touch the side labeled Y. √
- Raise your hand when you know how many units Y is. √
- How many units is Y? (Signal.) *5.*
- X is 4 and Y is 5. So 4 squared plus 5 squared equals H squared.
- Say that equation. (Signal.) $4^2 + 5^2 = H^2.$
- 4 squared is 16, and Y squared is 25. So H squared is 16 plus 25. That's 41.
- To find H, you'd have to take the square root of 41. That's about 6.4.

Students then work with triangles, apply the equation, and figure out $h^2$, not $h$.

Here's an example:

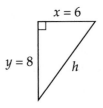

First, students write the general equation: $x^2 + y^2 = h^2.$

Next, they substitute for $x$ and $y$: $6^2 + 8^2 = h^2.$

Next, they rewrite the equation with values for $x^2$ and $y^2$: $36 + 64 = h^2.$

Next, they solve for $h^2$: $\boxed{100 = h^2}$

For these initial examples, $y$ is the vertical side, and $x$ is the horizontal side.

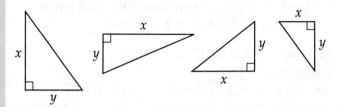

In lesson 88, students first learn the procedure for solving equations for $h$. Here's the explanation:

- Each problem shows H squared and what it equals. To work these problems, you take the square root of both sides.
b. The box shows an example.
- (Teacher reference:)

$$121 = h^2$$
$$\sqrt{121} = \sqrt{h^2}$$
$$11 = h$$

- If $H^2$ equals 121, you take the square root of both sides to figure out what H equals. You take the square root of $H^2$. That's H.
- You also take the square root of 121. That's 11.
- So if $H^2 = 121$, H = 11.

In the same lesson, students apply the procedure to triangles that are shown in different orientations.

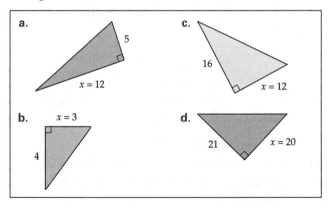

For these examples, one of the shorter sides is arbitrarily labeled $x$.

**Teaching note:** Unit names are not shown for these triangles and are not introduced until lesson 90.

In lesson 91, students solve for $x$ or $y$. Triangles are shown in different orientations and the labels for $x$ and $y$ are arbitrary. Here's the introduction:

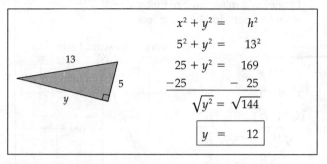

$$x^2 + y^2 = h^2$$
$$5^2 + y^2 = 13^2$$
$$25 + y^2 = 169$$
$$-25 \qquad -25$$
$$\sqrt{y^2} = \sqrt{144}$$
$$y = 12$$

b. The box shows a new kind of problem. The value for H is shown, but the length of Y is not shown.
- You find the missing value by starting with the equation X squared plus Y squared equals H squared.
- Then you put in the values the triangle gives: $5^2 + Y^2 = 13^2$. That's $25 + Y^2 = 169$.
- We subtract 25 from each side. $Y^2 = 144$.
- You take the square root of both sides.
- If $Y^2 = 144$, Y equals the square root of 144.
- What number is that? (Signal.) *12.* Yes, Y is 12 units long.

Word problems start in lesson 94. Some require more than one calculation.

Here's an example:

a. Jan starts at point A, then goes to point B and point C. Hal starts at point A and goes directly to point C. How much shorter is Hal's route than Jan's route?

Students apply the convention of identifying the vertical side of the triangle as y. Students start with the general equation, find the length of the hypotenuse, add the length of $x$ and $y$ to find Jan's route, and subtract $h$ to find the difference between the two routes.

To work the problems presented in the next lesson (95), students make diagrams and label the sides.

Here's an example:

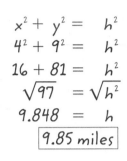

**a.** Mary lives 9 miles north of a train station. Rick lives 4 miles west of that station. How far apart do Mary and Rick live?

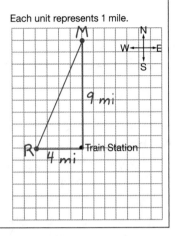

Each unit represents 1 mile.

$$x^2 + y^2 = h^2$$
$$4^2 + 9^2 = h^2$$
$$16 + 81 = h^2$$
$$\sqrt{97} = \sqrt{h^2}$$
$$9.848 = h$$
$$\boxed{9.85 \text{ miles}}$$

Students construct a triangle: $y$ is 9 and $x$ is 4. They find the hypotenuse and write the answer (9.85 miles).

After working problems of this type in lessons 95–97, students work problems that tell about objects moving in the same direction or opposite directions from the same starting point. Students learn the rule that if things move in the same direction, you subtract to figure out how far apart they are; if things move in opposite directions, you add to figure out how far apart they are.

Here are examples of these problem types:

*Tom walks 12 miles east from the school. Jerry walks 7 miles west from the school. How far apart are the boys?*

*Hilary and Barbara started the race together. After an hour, Hilary had run 12.3 kilometers. Barbara had run 10.8 kilometers. How far apart were the 2 runners after the hour?*

Later, students work problems that tell how far apart objects are and figure out how far one of them is from the starting point. Here are examples of those types presented in lesson 99:

*Two runners start from the same point. Runner A goes east 6.5 miles. Runner B goes west until they are 14.3 miles apart. How far from the starting point is runner B?*

*Two snails leave from the same point. The first snail goes north 12 feet. The second goes east until the 2 snails are 15 feet apart. How far east of the starting point is the second snail?*

*Two goats race downhill. They continue until the gray goat is 540 yards ahead of the black goat. If the black goat had gone 1611 yards, how far had the gray goat gone?*

## SURFACE AREA, COMPLEX AREA (Lessons 113–117)

Starting in lesson 113, students work problems that involve the surface area of boxes. The procedure they follow is to compute the area of one side, one end, and either the top or the bottom. Then they double the total of these sides.

In lesson 115, students work problems that have two-dimensional shapes with cutout parts. Students figure out the area of the entire object and the area of the cutout part, then subtract to find the area of the entire object with the part cut out.

Here are two examples:

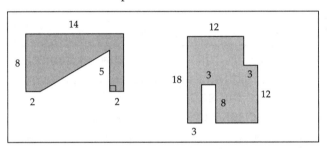

Finally, students learn to find the area of a trapezoid by constructing right triangles needed to convert the trapezoid into a rectangle. Students compute the area of the rectangle and subtract the areas of the triangles.

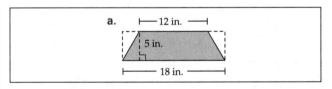

## VOLUME (Lessons 117–118)

In lesson 117, students learn that the volume of a three-dimensional figure is the area of a face times the other dimension.

One face of the figure is shaded.

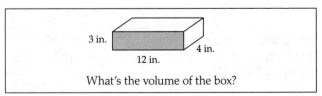

What's the volume of the box?

In lesson 117, students use the equation $V = (Af)\,d$ to find the volume or one of the dimensions (height, width, depth) of rectangular prisms.

In lesson 118, students use the equation $V = (Af)\,d$ to find the volume of a variety of prisms.

Here are two examples:

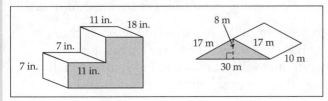

The geometry track covers a great deal of information. Most of the material is reviewed and used frequently enough, however, for students to retain it.

# Exponents (Lessons 46–116)

Students learn to express repeated multiplication of a value as a base and as an exponent. They also learn to write the repeated multiplication for a base and an exponent. Students learn to write fractions with repeated multiplication in the denominator and to simplify fractions that have repeated multiplication in both numerator and denominator.

In lesson 57, students learn to express given values with either a positive or a negative exponent. In lesson 65, students begin work on combining exponents. Negative bases are introduced in lesson 109. In lesson 112, students learn to raise a power to a power. Finally, students learn to substitute values that have a base and an exponent.

The groundwork for exponents occurs in lessons 46–49. Students learn the terms *base* and *exponent*. They rewrite repeated multiplication of letters or numbers as bases and exponents. They also translate a base and an exponent as repeated multiplication. Here's part of the exercise from lesson 47:

> **a.** $6 \times 6 \times 6 =$ ▓

a. You've worked with repeated multiplication of the same value. What do we call the value that is repeated? (Signal.) *The base [number].*
- What do we call the number for how many times it's repeated? (Signal.) *The exponent.*
- (Repeat step a until firm.)

b. Listen: $5 \times 5 \times 5$. What's the base? (Signal.) *5.*
- What's the exponent? (Signal.) *3.*
- (Write on the board:)                    [47:5A]

$$5^3$$

- Here's what you write for $5 \times 5 \times 5$.
- Everybody, read this value. (Signal.) *5 to the 3rd.*
c. (Write to show:)                    [47:5B]

$$5^3$$
$$4^5$$

- Everybody, read this value. (Signal.) *4 to the 5th.*
- Some of these problems show a base number and an exponent. Other problems show the multiplication. You'll write complete equations. You'll copy what's shown and write what goes on the other side of the equal sign.
b. Problem A. The multiplication is shown. You'll show the base and exponent. What base is shown? (Signal.) *6.*
- How many times is 6 shown? (Signal.) *3.*
- What will you write to complete the equation? (Signal.) *6 to the 3rd.*
- Write the complete equation for A. Pencils down when you're finished. √
- (Write on the board:)                    [47:5C]

**a.**    $6 \times 6 \times 6 = 6^3$

- Here's what you should have. What does 6 x 6 x 6 equal? (Signal.) *6 to the 3rd.*

In lessons 50 and 51, students learn that different groups of the same multiplied value may be expressed as bases with exponents. They learn that regardless of the grouping, the sum of the exponents is the same.

Here's part of the exercise from lesson 50:

**a.** $(8 \times 8) \times (8 \times 8 \times 8) \times (8 \times 8)$

$8^7 =$ ▮ $\times$ ▮ $\times$ ▮

- For each item, you'll write the complete equation with exponents.
b. Problem A. The multiplication shows 8 seven times.
- Say the base and exponent for all the 8s. (Signal.) *$8^7$*.
  So no matter how the 8s are multiplied together, the exponents must add up to 7.
- You can see the groups set off with parentheses.
- Touch the first group. √
  Tell me the base and exponent you'll write for the first group. (Signal.) *$8^2$*.
- Next group.
  Tell me the base and exponent. (Signal.) *$8^3$*.
- Last group.
- Tell me the base and exponent. (Signal.) *$8^2$*.
- The exponents are 2 and 3 and 2. Do the exponents add up to 7? (Signal.) *Yes.*
- So the whole equation is $8^7 = 8^2 \times 8^3 \times 8^2$.
c. Say the equation. (Signal.) *$8^7 = 8^2 \times 8^3 \times 8^2$.*
- Write that equation. Pencils down when you're finished. √
- (Write on the board:)  [50:1A]

**a.**   $8^7 = 8^2 \times 8^3 \times 8^2$

- Here's what you should have.

Additional work with different groups shows that no matter how the bases are grouped, the sum of the exponents remains the same, which means that it is possible to add the exponents from the subgroups to obtain the exponent for the total group.

In lessons 52–54, students work with fractions that have the same base multiplied in both numerator and denominator. To show what the fraction equals, students answer two questions:

- ***Is the base shown more times in the numerator or denominator?***

- ***How many more times is the base shown?***

The answer to the first question tells whether the base will be written in the numerator or denominator. The answer to the second equation tells what the exponent is.

Here's an example from lesson 52:

**a.** $\dfrac{5}{5 \times 5 \times 5 \times 5}$

b. Fraction A. What's the base? (Signal.) 5.
- Is the base shown more times in the numerator or in the denominator? (Signal.) *Denominator.*
- So you'll write 5 in the denominator.
- How many more times does the base appear in the denominator? (Signal.) *3.*
- So what's the exponent? (Signal.) *3.*
  Yes, fraction A equals 1 over $5^3$.
- What does it equal? (Signal.) *1 over $5^3$.*
- (Write on the board:)  [52:3A]

**a.**   $\dfrac{1}{5^3}$

- You write 1 in the numerator and $5^3$ in the denominator.
- Write the fraction for A. Pencils down when you're finished. √

Students work with other examples, some of which show the base multiplied more times in the numerator than in the denominator. For these, students do not show 1 in the denominator.

Negative exponents are introduced in lesson 57. Students learn that a base and an exponent in either the numerator or the denominator may be moved to the opposite part of the fraction and shown with the opposite sign.

Here's part of the introduction:

---

♦ Move the **base** to the **opposite part** of the fraction.
♦ Change the **sign of the exponent** to the **opposite sign**.

---

| | | |
|---|---|---|
| $\frac{1}{6^4} = 6^{-4}$ | $6^3 = \frac{1}{6^{-3}}$ | $\frac{1}{6^{-1}} = 6^1$ |

b. The box shows how you can rewrite any base with an exponent: You move the **base** to the **opposite part** of the fraction. You change the **sign of the exponent** to the **opposite sign**.

c. The first thing you do is move the base to the opposite part of the fraction.
• What's the first thing you do? (Signal.) *Move the base to the opposite part of the fraction.*
• If the base is shown in the denominator, where does it move to? (Signal.) *The numerator.*
• If the base is shown in the numerator, where does it move to? (Signal.) *The denominator.*
• The next thing you do is change the sign of the exponent to the opposite sign.
• What's the next thing you do? (Signal.) *Change the sign of the exponent to the opposite sign.*
• If the exponent is positive, what do you change it to? (Signal.) *Negative.*
• If the exponent is negative, what do you change it to? (Signal.) *Positive.*
• (Repeat step c until firm.)

d. Remember to move the base to the opposite part, and change the **sign** of the exponent.

e. These equations show how to rewrite values with positive exponents as values with negative exponents. The first equation shows the fraction 1 over 6 to the 4ᵗʰ. That equals 6 to the minus 4ᵗʰ.
• What does 1 over 6 to the 4ᵗʰ equal? (Signal.) *6 to the minus 4ᵗʰ.*

f. The next equation shows 6 to the 3ʳᵈ. What does 6 to the 3ʳᵈ equal? (Signal.) *1 over 6 to the minus 3ʳᵈ.*

g. The last equation shows 1 over 6 to the minus 1ˢᵗ. What does 1 over 6 to the minus 1ˢᵗ equal? (Signal.) *6 to the 1ˢᵗ.*

In subsequent lessons, students manipulate fractions that have different bases multiplied in the numerator and denominator. For example:

$$\frac{3^{-2}}{d^{-2}\,5^4}, \quad \frac{m^{-2}\,n^{-3}\,4^3}{f^5}, \quad \frac{d^5\,r^{-2}\,c}{2^{-3}z^4}$$

Students perform different manipulations, such as writing all the values in the numerator or writing all the values with positive exponents.

Starting with lesson 65, students learn to simplify expressions by combining exponents of multiplied values that have the same base. For instance:

---

**a.** $m^{-3}\,4^2\,m^8\,4^3$

---

b. Item A. $M^{-3}\,4^2\,M^8\,4^3$.
• What are the 2 bases? (Signal.) *M and 4.*
• Are the bases multiplied together? (Signal.) *Yes.*
• So you can simplify the expression. Say the problem for combining the exponents of M. Get ready. (Signal.) $-3 + 8$.
• So what exponent do you write for the base of M? (Signal.) *5.*
• Say the problem for combining the exponents of 4. Get ready. (Signal.) $2 + 3$.
• So what exponent do you write for the base of 4? (Signal.) *5.*
• Copy expression A and write an equation to show the simplified expression. Pencils down when you're finished.
(Observe students and give feedback.)
• (Write on the board:)                    [65:3D]

**a.**      $m^{-3}\,4^2\,m^8\,4^3 = m^5\,4^5$

• Here's what you should have. Read the simplified expression. (Signal.) $M^5\,4^5$.

Starting with lesson 69, students learn to express the number base as a value without an exponent. For the example above, the final simplified value would be $1024m^5$.

Starting in lesson 77, students work problems that have expressions with the same bases shown in both numerator and denominator. Students move all values to the numerator, then combine exponents.

Here's a problem and the solution from lesson 78:

$$\frac{3k^4 r^{-1}}{k^2 r^3 3^5} = 3k^4 r^{-1} k^{-2} r^{-3} 3^{-5}$$

$$= 3^{-4} k^2 r^{-4}$$

$$= \boxed{\frac{1}{81} k^2 r^{-4}}$$

In lesson 106, students learn to discriminate between bases that are multiplied and those that are combined. Here's part of the presentation from lesson 106:

> **a.** $r^4 \times r$ **b.** $r^4 + 5r^4$

- For some of these problems, you'll multiply. For others, you'll combine.
b. Remember the rules:
- If you multiply **bases** that are the same, you combine the exponents. Say that rule. (Signal.) *If you multiply bases that are the same, you combine the exponents.*
- If you combine **terms,** the base and exponent must be the same. Say that rule. (Signal.) *If you combine terms, the base and the exponents must be the same.*
c. Problem A: R⁴ times R.
- Are you multiplying or combining the terms? (Signal.) *Multiplying.*
- So you combine the exponents. That's 4 plus 1.
d. Problem B: R⁴+ 5R⁴.
- Are you multiplying or combining the terms? (Signal.) *Combining.*
- Are the bases and exponents the same? (Signal.) *Yes.*
  So you can combine the like terms.
- How many Rs to the fourth are there? (Signal.) *6.*
e. Work problems A and B. Pencils down when you're finished.
  (Observe students and give feedback.)
- (Write on the board:)                [106:3A]

> **a.** $r^4 \times r = \boxed{r^5}$
>
> **b.** $r^4 + 5r^4 = \boxed{6r^4}$

- Here's what you should have.

Students work with expressions that involve both multiplication and combining like terms. For example:

$$(3m^2 \times 2m^2) - 2m^4$$

Negative bases are introduced in lesson 109. Students learn rules for values that have even-numbered exponents and odd-numbered exponents.

Here's a part of the introduction:

> - If the exponent is an **odd** number, the sign in the answer is **minus.**
>
> - If the exponent is an **even** number, the sign in the answer is **plus.**
>
> ---
>
> $\overset{+}{(-a)}\ \overset{+}{(-a)} \cdot \overset{+}{(-a)}\ (-a) \cdot \overset{+}{(-a)}\ (-a) = \boxed{+a^6}$
>
> ---
>
> $\overset{+}{(-a)}\ \overset{+}{(-a)} \cdot \overset{+}{(-a)}\ (-a) \cdot \overset{-}{(-a)} = \boxed{-a^5}$

- The box gives rules about bases that have a minus sign.
  If the exponent is an **odd** number, the sign in the answer is **minus.** If the exponent is an **even** number, the sign in the answer is **plus.**
b. So if the base has a minus sign and the exponent is an **odd** number, the sign in the answer is **minus.**
- Once more: If the base has a minus sign and the exponent is odd, what's the sign in the answer? (Signal.) *Minus.*
c. And if the base has a minus sign and the exponent is an **even** number, the sign in the answer is a **plus.**
- Once more: If the base has a minus sign and the exponent is even, what's the sign in the answer? (Signal.) *Plus.*
- The reason is that a **minus** times a **minus** is a **plus.**
- So each pair of negative bases that are multiplied give a positive value. If the exponent is 6, you have three pairs of negative values. So the answer is positive.

- (Teacher reference:)

$$\overset{\lceil\ +\ \rceil}{(-a)\ (-a)} \cdot \overset{\lceil\ +\ \rceil}{(-a)\ (-a)} \cdot \overset{\lceil\ +\ \rceil}{(-a)\ (-a)} = \boxed{+a^6}$$

$$\overset{\lceil\ +\ \rceil}{(-a)\ (-a)} \cdot \overset{\lceil\ +\ \rceil}{(-a)\ (-a)} \cdot \overset{\lceil\ -\ \rceil}{(-a)} \quad = \boxed{-a^5}$$

d. Look at the box that shows 6 negative bases. The sign of each pair is shown above. For the first 2, the sign is positive. Touch that sign. √
- For the next 2, the sign is positive.
- For the last 2, the sign is positive.
- Remember, if the exponent is an even number, the sign in the answer is plus. The answer is $+A^6$.

e. The next box shows 5 negative bases. You can see the sign above the first pair. What sign? (Signal.) *Plus.*
- What's the sign above the next pair? (Signal.) *Plus.*
- What's the sign above the fifth base? (Signal.) *Minus.*
- You're multiplying a positive times a negative. So the answer is negative. The answer is $-A^5$.

f. Listen: Minus $A^4$.
- Is the exponent even or odd? (Signal.) *Even.*
- So is the value positive or negative? (Signal.) *Positive.*

g. New exponent: Minus $A^{17}$.
- Is the exponent even or odd? (Signal.) *Odd.*
- So is the value positive or negative? (Signal.) *Negative.*

h. Listen: Minus $4B^9$.
- Is the exponent even or odd? (Signal.) *Odd.*
- So is the value positive or negative? (Signal.) *Negative.*

i. Listen: Minus 1 half $D^{10}$.
- Is the exponent even or odd? (Signal.) *Even.*
- So is the value positive or negative? (Signal.) *Positive.*

j. Listen: Minus $6K^{13}$.
- Is the exponent even or odd? (Signal.) *Odd.*
- So is the value positive or negative? (Signal.) *Negative.*

k. Listen: Minus $8Q^{16}$.
- Is the exponent even or odd? (Signal.) *Even.*
- So is the value positive or negative? (Signal.) *Positive.*

- (Repeat steps f–k until firm.)

**Teaching note:** Students have learned that everything that is multiplied inside a set of parentheses is raised to the power of the exponent outside the parentheses.

In the next lesson, students work simplification problems that require two steps. They first rewrite the expression as a base with a single exponent, then simplify.

For example:

$$(-2)^2\ (-2)^4 = (-2)^6 = +64$$

In lesson 111, students discriminate between values that are shown with parentheses and those not shown with parentheses. For example:

$$(2r)^3 \text{ and } 2r^3$$

If there are no parentheses, only the letter or number that has the exponent is raised to the specified power. Students also learn to rewrite complex bases as a base with an exponent raised to a power. For example:

$$5a^2 \times 5a^2 \times 5a^2 = (5a^2)^3$$

Finally, students learn that if a base with an exponent is raised to a power, multiplying the exponent inside the parentheses times the exponent outside gives the simplified value,

$$(5a^2)^3 = 5^3a^6 = 125a^6.$$

Students apply this operation to equations that require substitution, such as,

$$r = 4^2 \qquad\qquad m = 4^2$$
$$r^5 = \blacksquare \qquad\qquad m^{\blacksquare} = 4^{12}$$

## Simultaneous Equations (Lessons 73–112)

This track introduces two methods of solving simultaneous equations. The first is to combine a pair of equations to eliminate one of the unknowns. The other is to solve for an unknown in one equation and substitute that value in the other equation.

The first problem type presents two equations with a single unknown. Students learn the rule that if both equations are true, the combined equation is true. Students first combine these equations, then solve for the unknown. Here's part of the introduction from lesson 73:

$$
\begin{array}{rcr}
2t + 4 &=& -2 \\
t - 1 &=& -4 \\
\hline
3t + 3 &=& -6
\end{array}
$$

d. The next box shows 2 equations with a letter term and 2 number terms: 2T + 4 = − 2 and T − 1 = − 4.
- We'll combine the T terms and then each of the number terms.
- Say the problem for combining the T terms. (Signal.) *2T + T.*
- What's the answer? (Signal.) *3T.*
- Say the problem for combining the next terms. (Signal.) *+ 4 − 1.*
- What's the answer? (Signal.) *+ 3.* Yes, you need the plus sign.
- Say the problem for the last terms. (Signal.) *− 2 − 4.*
- What's the answer? (Signal.) *− 6.*
- Read the combined equation. (Signal.) *3T + 3 = − 6.*
e. If both equations are true, the combined equation is also true. Solve that equation for T. Pencils down when you're finished. (Observe students and give feedback.)

- Check your work.
- What does T equal? (Signal.) *− 3.*
- If T equals − 3 in the combined equation, it equals − 3 in **both** of the **original** equations.
- Replace T with − 3 in the top equation. Remove the parentheses. Pencils down when you've done that much. (Observe students and give feedback.)
- Check your work.
- (Write on the board:) [73:1A]

$$
\begin{array}{r}
2\ t + 4 = -2 \\
2\,(-3) + 4 = -2 \\
-6 + 4 = -2
\end{array}
$$

- Here's what you should have. 2T is 2 times − 3. That's − 6.
- Raise your hand when you know what you get when you combine − 6 + 4. √
- Everybody, what do you get? (Signal.) *− 2.*
- So the equation is true.
- Now replace T with − 3 in the second equation. Pencils down when you've written the equation with a value for T. (Observe students and give feedback.)
- Read the equation you get. (Signal.) *− 3 − 1 = − 4.* That's true, too.
f. Remember, if the equations you combine are true, the combined equation is also true.

In lesson 73, students work other problems of the type shown. They solve for the unknown, then substitute in each equation to test whether the equations are true.

In lesson 75, students work with pairs of equations that have the same two unknowns. These equations are designed so that when the equations are combined, one of the letters is eliminated.

For example:

$$
\begin{array}{rcr}
n + 2t &=& 20 \\
2n - 2t &=& 10 \\
\hline
3n &=& 30.
\end{array}
$$

The combined equation is $3n = 30$.

Students solve for $n$, then substitute and solve for $t$.

Starting with lesson 79, students work with pairs of equations such as these:

$$-5f = 10$$
$$2r - f = 14$$

Students eliminate the letter that appears in both equations. To do that, they must multiply one of the equations by a value that yields a letter term with the opposite sign and same number as the letter term in the other equation. For the example above, students multiply the second equation by $-5$. To figure out what to multiply by, they work the problem $-f(\ ) = 5f$. The signs of $-f$ and $5f$ are different, so the sign for the missing value is minus. $f$ times $5 = 5f$, so the missing value is $-5$. (Note that students have worked on this subskill in isolation since lesson 74.)

Here's the solution for the problem above:

$$-5\,(2r - f) = 14\,(-5)$$
$$-10r + 5f = -70$$
$$-5f = 10$$
$$\overline{\phantom{-10r}}$$
$$-10r = -60$$
$$\left(\tfrac{1}{10}\right)10r = 60\left(\tfrac{1}{10}\right)$$

$$\boxed{r = 6}$$

*Teaching note:* To work these problems, students do not have to decide which unknown to eliminate because only one unknown appears in both equations. For the types that follow, students must decide which letter is easier to eliminate.

The next problem type is introduced in lesson 82. The key part to this problem type is identifying the pair of letter terms that are multiples.

Here's part of the exercise from lesson 82:

> **a.** $2g + 2r = -4$
>
> $-5g + 6r = 32$

• Each item has 2 pairs of letter terms.
b. Item A: $2G + 2R = -4$, and $-5G + 6R = 32$.
• There are G terms and R terms.
• Everybody, read the G terms. (Signal.) *2G, −5G.*
• Read the R terms. (Signal.) *+ 2R, + 6R.*
• You're going to eliminate 1 of these letters by multiplying.
• The numbers for that pair of terms are multiples. What letter has numbers that are multiples? (Signal.) *R.*
• Touch the term you'll change. √
• Which term will you change? (Signal.) *2R.*
• What will you change 2R into? (Signal.) *− 6R.*
• (Repeat until firm.)

*Teaching note:* In step b above, students may not identify the pair of terms that are multiples. To correct, ask about the specific values:

• The numbers for the G terms are 2 and 5.

• Is 5 a multiple of 2?

• The numbers for the R term are 2 and 6.

• Is 6 a multiple of 2?

• So you're going to eliminate R.

To solve the problem, students multiply the first equation by minus 3; then they combine the resulting equation with the second equation.

**a.** 
$$-3\,(2g + 2r) = -4\,(-3)$$
$$-6g - 6r = 12$$
$$-5g + 6r = 32$$
$$\overline{\phantom{-11g}}$$
$$-11g = 44$$
$$\left(\tfrac{1}{11}\right)11g = -44\left(\tfrac{1}{11}\right)$$

$$\boxed{g = -4}$$

In lesson 83, students solve for both letters. After solving for one letter, they substitute that value in the equation they did not change and solve for the other letter.

In lesson 87, students learn that they are able to solve simultaneous equations by solving one equation for a letter and then substituting that value in the other equation.

| a. $-1 + 2r = -3j$ | b. $3r - p = 7$ |
|---|---|
| $10 - 5r = 9j$ | $-4r = -8$ |

- These are equation pairs. Some of these problems have 2 letters in **both** equations. Some have only 1 letter in 1 of the equations.
- b. Problem A. Look at the letters. √ Do both equations have 2 unknowns? (Signal.) *Yes.*
- c. Problem B. Look at the unknowns. √ Do both equations have 2 unknowns? (Signal.) *No.*
- Read the equation with 1 unknown. (Signal.) *– 4R = – 8.*
- So you can work problem B a simple way. First solve for R. Then substitute for R in the other equation.
- Copy the equation $-4R = -8$ and solve it. Pencils down when you've done that much. √
- Everybody, what does R equal? (Signal.) *2.*
- Substitute for R in the other equation and solve it for P. Pencils down when you're finished. (Observe students and give feedback.)
- Everybody, what does P equal? (Signal.) *– 1.*

In lesson 107, students work with simultaneous equations on the coordinate system. Students solve a pair of equations for both $x$ and $y$. They plot that solution point, then plot both lines and check whether the lines intersect at that point.

Here's part of the exercise that follows the introduction:

| a. $y = 3x - 5$ |
|---|
| $y = -2x + 5$ |

- b. Each item shows a pair of equations for lines that intersect.
- c. Item A. You'll solve the pair of equations for X and for Y. You'll multiply one of the equations, then combine equations to eliminate a letter. Which letter will you eliminate? (Signal.) *Y.* Yes, Y.
- Multiply one of the equations by –1. Then solve for X and for Y. Pencils down when you've done that much. (Observe students and give feedback.)
- Check your work.
- Everybody, what does X equal? (Signal.) *2.*
- What does Y equal? (Signal.) *1.*
- d. If the solution is correct, the lines for these equations will intersect at 2 comma 1.
- Plot that point on coordinate system A. Label it 2 comma 1. √
- e. Now plot the line for the equation $y = 3x - 5$. That line should go through the point 2 comma 1. (Observe students and give feedback.)
- f. Now plot the line for the equation $Y = -2X + 5$. That line should also go through 2 comma 1. (Observe students and give feedback.)
- It works!

Lesson 109 introduces substitution with a pair of equations, each with two unknowns. One of the equations is already solved for a letter. Students simply substitute the value for that letter in the other equation. Here's an example:

$$p = 2 + r$$
$$5r - 3p = -12$$

The first equation tells what P equals. Students substitute for P in the other equation. The result is an equation with only one unknown:

$$5r - 3(2 + r) = -12$$

Students distribute and solve for R.

$$5r - 6 - 3r = -12$$

$$2r - 6 \quad\quad = -12$$

$$\underline{+ 6 \quad\quad\quad + 6}$$

$$\left(\tfrac{1}{2}\right) 2r \quad\quad = -6\left(\tfrac{1}{2}\right)$$

$$\boxed{r \quad\quad\quad = -3}$$

Then they substitute minus 3 for R in either equation and solve for P.

The final problem type in the track involves word problems that generate a pair of simultaneous equations.

Here's an example from lesson 111:

> *The number of girls is 3 less than twice the number of boys. The total number of children is 24. How many boys are there? How many girls are there?*

Students write an equation for the first sentence and for the second sentence:

$$g = 2b - 3$$

$$g + b = 24$$

Students substitute $2b - 3$ for $g$ in the second equation and solve for $b$. Then they substitute in either equation to solve for $g$.

**Teaching note:** By the time students reach this point in the program, they have various options. Solving the problem by substitution is arbitrary. They could solve the problem by combining the equations. Even when students solve by substitution, they have options about how they could write the second equation. There are 4 options:

$$g + b = 24$$

$$b + g = 24$$

$$24 = g + b$$

$$24 = b + g$$

## Probability (Lessons 84–110)

This track teaches students how to work different problem types involving trials. The four basic types ask about the composition of the set or the number of trials.

- *If there are a total of 14 cards and 3 of them are red, how many trials would you expect to take to draw 12 red cards from the deck?*

- *If you took 70 trials of drawing cards from the deck, on how many trials would you expect to draw a red card?*

- *If you took 150 trials at rolling a 6-sided die and a blue side was on top on 25 trials, how many blue sides does the die probably have?*

- *There are marbles in a bag. 3 are blue. If you take 80 trials and pull out 20 blue marbles, how many total marbles would you estimate are in the bag?*

Then students learn about the probability of more than one independent event occurring (such as drawing two white cards in succession from a deck if the first card is returned to the deck). They also learn the probability of one or another event occurring (such as drawing a white card or a yellow card from a deck).

Finally, students work problems in which the size of the group changes as items are removed, which means that the probability fraction changes for each trial.

> *There are 12 marbles in a bag. 3 are blue. What's the probability of drawing a blue marble on the first trial? If the marble is not returned to the bag, what's the probability of drawing a blue marble on the second trial?*

In lesson 84, students learn about writing probability fractions based on the composition of a set. They write the total number of objects in the set as the denominator and the number of a specified subtype as the numerator. If there are 9 objects in a bag and 4 of them are red, the basic probability fraction for a red object is 4/9.

Next, students learn how the numbers in the fraction relate to trials and expectations. The general strategy is for students to recognize that the number of objects not only gives the fraction for the objects but also the basic fraction for the number of trials one takes and the expected number of "winners." If the fraction for the object shows a total of 8 objects and 5 are red, the fraction for objects is 5/8. That's also the basic fraction for the number of trials and the number of times a red object would be drawn from the bag.

Here's part of the introduction:

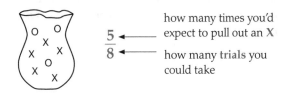

- ◆ The fraction that tells about your chances of pulling an X from a bag tells about **trials.**

- ◆ When you take a trial, you pull 1 object from the bag without looking. Then you return it to the bag and mix up the objects.

$$\frac{5}{8}$$ how many times you'd expect to pull out an **X**
how many **trials** you could take

b. I'll read the rule in the box: The fraction that tells about your chances of pulling an X from a bag tells about **trials.**
- What does it tell about? (Signal.) *Trials.*
- When you take a trial, you pull 1 object from the bag without looking. Then you return it to the bag and mix up the objects.
- If the bag starts out with 8 objects in it, there would always be the same 8 objects for each trial. They would just be mixed up different ways.
- Remember, no matter how many trials you take, the bag always has the same number of objects in it.
c. Touch the bag in part 3. √
- Everybody, what's the fraction for that bag? (Signal.) *5/8.*
- You can use that fraction to tell how many trials you could take and how many times you would expect to pull out an X.
- Listen again: You can use that fraction to tell how many trials you could take and how many times you would expect to pull out an X.

d. The denominator tells how many trials you could take. The numerator tells how many times you'd expect to pull out an X.
- What does the **denominator** tell? (Signal.) *How many trials you could take.*
- What does the **numerator** tell? (Signal.) *How many times you'd expect to pull out an X.*
- (Repeat step d until firm.)
e. The denominator is 8, so you could take 8 trials. You'd expect to pull out an X 5 times.
f. Listen: If the fraction is 7/8, how many trials could you take? (Signal.) *8.*
- How many times would you expect to pull out an X? (Signal.) *7.*
- If the fraction is 4/11, how many trials could you take? (Signal.) *11.*
- How many times would you expect to pull out an X? (Signal.) *4.*
- (Repeat step f until firm.)

In lesson 85, students learn the basic equation form for probability. One fraction tells about the object; the other tells about trials.

Here's part of the introduction:

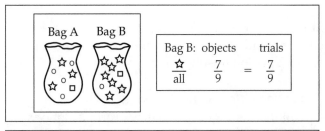

Bag A   Bag B

Bag B:   objects   trials
$$\frac{☆}{all} \quad \frac{7}{9} = \frac{7}{9}$$

a. A person took 15 trials and pulled out a star on 6 of those trials.

objects   trials
$$\frac{☆}{all} \quad \boxed{\phantom{x}} = \boxed{\phantom{x}}$$

e. The next box shows the simple equation that you can write for objects and trials based on bag B.
- The simple equation is 7/9 = 7/9.
- The first fraction tells about the objects.
f. What does the first fraction tell about? (Signal.) *The objects.*
- What does the second fraction tell about? (Signal.) *The trials.*
- (Repeat step f until firm.)

g. Yes, there are 9 objects in the bag, and 7 of them are stars. That means if you took 9 trials, you would expect to pull out a star on 7 of those trials.

h. I'll say statements. You'll tell me if they tell about the number of objects in the bag or about the number of trials.

• A person took 7 trials and pulled out a star on 6 of those trials. Does that statement give numbers for the objects or numbers for the trials? (Signal.) *Trials.*

• There were 7 objects in a bag. 6 of them were stars.
  Does that statement give numbers for the objects or numbers for the trials? (Signal.) *Objects.*

b. For each statement, you'll write a fraction. If it tells about objects, you'll write it under **objects.** If it tells about trials, you'll write it under **trials.**

c. Listen to statement A: A person took 15 trials and pulled out a star on 6 of those trials.

• Write that fraction where it belongs. √

• Everybody, what's the fraction you wrote? (Signal.) *6/15.*

• Is that a fraction for the objects or for the trials? (Signal.) *Trials.*

• (Write on the board:)                    [85:1A]

| a. | objects | trials |
|---|---|---|
| $\frac{☆}{all}$ | $\square$ = | $\frac{6}{15}$ |

• Here's what you should have. 6 fifteenths is the fraction. It gives numbers for trials.

Students write fractions in the appropriate place for four more items.

*Teaching note:* This orientation seems quite simple, but it is important because students often have serious difficulties in understanding where the numbers from word problems are to go in the equation. If students are firm in their understanding that the first fraction tells about the object (or setup) and that the other fraction tells about trials and expectations, they won't make mistakes.

In lesson 86, students learn that any fraction equivalent to the probability fraction based on the set tells about trials and expectations.

Here's part of the introduction from lesson 86:

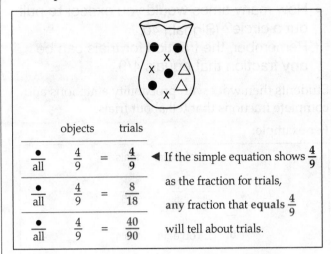

| | objects | trials | |
|---|---|---|---|
| $\frac{●}{all}$ | $\frac{4}{9}$ | = | $\frac{4}{9}$ | ◄ If the simple equation shows $\frac{4}{9}$ |
| $\frac{●}{all}$ | $\frac{4}{9}$ | = | $\frac{8}{18}$ | as the fraction for trials, any fraction that **equals** $\frac{4}{9}$ |
| $\frac{●}{all}$ | $\frac{4}{9}$ | = | $\frac{40}{90}$ | will tell about trials. |

b. You've learned that you can make predictions based on objects. If there are 9 objects in a bag and 4 of them are circles, you can write a fraction for the bag.

• Everybody, what fraction? (Signal.) *4/9.*

• And you can predict that if a person took 9 trials, the person would probably pull a circle from the bag 4 times.

• The first box shows the simple equation for that prediction: 4/9 = 4/9. If you used the fraction based on the objects, how many trials would you take? (Signal.) *9.*

• How many times would you expect to pull a circle from the bag? (Signal.) *4.*

c. I'll read the rule in the next box: If the simple equation shows 4/9 as the fraction for trials, **any** fraction that equals 4/9 will tell about trials.

d. The second equation tells about the same bag but has a different fraction for the trials. That fraction equals 4/9.

• What's the fraction for the trials? (Signal.) *8/18.*

• For that fraction, how many trials would you take? (Signal.) *18.*

• How many times would you expect to pull out a circle? (Signal.) *8.*

e. The last equation shows a different fraction for trials.

- What's the fraction for the trials? (Signal.) *40/90.*
- For that fraction, how many trials would you take? (Signal.) *90.*
- How many times would you expect to pull out a circle? (Signal.) *40.*
  f. Remember, the fraction for trials can be **any** fraction that equals 4/9.

Students then work with probability equations and complete fractions that tell about trials.

For example:

| | a. | objects | trials | |
|---|---|---|---|---|
| | | $\frac{\star}{\text{all}}$ | $\frac{3}{5}$ = | $\frac{\phantom{25}}{25}$ |

| | b. | objects | trials | |
|---|---|---|---|---|
| | | $\frac{\star}{\text{all}}$ | $\frac{2}{3}$ = | $\frac{8}{\phantom{8}}$ |

In lesson 87, the set of "objects" is increased to include spinners, dice, and decks of cards.

Word problems are introduced in lesson 88.

| | object | trials |
|---|---|---|
| A die had 6 sides. 2 sides are black, and 4 sides are red. If a person rolled the die 15 times, how many times would you expect a **red side** to be on top? | $\frac{4}{6}$ = | $\frac{r}{15}$ |

b. The box shows a new kind of probability problem: A die has 6 sides. 2 sides are black, and 4 sides are red. If a person rolled the die 15 times, how many times would you expect a **red side** to be on top?
- The question names the parts you're interested in. What color are these parts? (Signal.) *Red.*
- So your fraction for the object will tell about the red parts and all the parts.
- You can see that fraction is 4/6.
c. The other fraction will tell about trials. Each roll of the die is a trial. So there are 15 trials.
- Is that the numerator or the denominator of the fraction? (Signal.) *Denominator.*
  Yes, 15 tells about all the trials, so it's in the denominator.

- We use the letter R for the number of times we expect red to be on top.
d. So the complete equation is 4/6 = R over 15.
- Copy the equation and solve for R. Pencils down when you've done that much. **(Observe students and give feedback.)**
- (Write on the board:) [88:3A]

$$(15)\,\frac{4}{6} = \frac{r}{15}\,(15)$$
$$10 = r$$

- Everybody, what does R equal? (Signal.) *10.*
- So you'd expect the red side to be on top for **10** of the 15 trials. The answer to the question is **10 trials.**
- (Write to show:) [88:3B]

$$(15)\,\frac{4}{6} = \frac{r}{15}\,(15)$$
$$10 = r$$
$$\boxed{10 \text{ trials}}$$

- Copy the answer and box it. √

**Teaching note:** Some of the problems that students work have the unknown in the denominator. Here's an example:

*There are 60 cards in a deck. 11 are red. 16 are blue. 27 are yellow. 6 are white. If a person pulls a blue card from the deck on 12 trials, how many total trials would you expect the person to take?*

Students write the equation $\frac{16}{60} = \frac{12}{t}$.

In lessons 80–83, students learned that the simplest way to work this kind of problem is to flip the equation so the unknown is on top: $\frac{60}{16} = \frac{t}{12}$.

The exercise in lesson 91 introduces problems that ask about a missing number in the fraction that tells about the object.

Here's part of the exercise:

**a.** A person takes 28 trials at drawing cards from a deck. The deck has 8 cards. On 7 trials, the person draws a card that is red. How many red cards would you expect to be in the deck?

$$\underset{\text{object}}{\dfrac{r}{8}} = \underset{\text{trials}}{\dfrac{7}{28}}$$

**b.** A die has 4 sides that are yellow. A person rolls a die 15 times. A yellow side is on top on 12 of those times. How many sides would you expect the die to have?

$$\underset{\text{object}}{\dfrac{4}{s}} = \underset{\text{trials}}{\dfrac{12}{15}}$$

**b.** The box shows probability problems that ask about the object, not the trials.

**c.** Problem A: A person takes 28 trials at drawing cards from a deck. The deck has 8 cards. On 7 trials, the person draws a card that is red. How many red cards would you expect to be in the deck?

- (Teacher reference:)

$$\underset{\text{object}}{\dfrac{r}{8}} = \underset{\text{trials}}{\dfrac{7}{28}}$$

- The problem asks about the red cards in the deck. That's **R**. The total number of cards in the deck is 8. So the fraction for the object is R over 8.
- The fraction for the trials is 7/28. The person takes 28 trials. On 7 of those trials the person draws a red card.
- To work the problem, you solve for R.

**d.** Problem B asks about the total number of parts in the object: A die has 4 sides that are yellow. A person rolls a die 15 times. A yellow side is on top on 12 of those times. How many sides would you expect the die to have?

- (Teacher reference:)

$$\underset{\text{object}}{\dfrac{4}{s}} = \underset{\text{trials}}{\dfrac{12}{15}}$$

- You can see the equation. The fraction for the object has 4 in the numerator because 4 sides are yellow. The denominator is S because we have to figure out the number of **sides** the die has.
- The fraction for the trials is 12/15. That means that on 12 of the 15 trials, a yellow side is on top. To work the problem, you flip the equation and solve for S.

In lesson 103, students learn about the probabilities of two or more events occurring on a single trial.

For this exercise, students learn that a trial may consist of more than one action. If the problem asks about drawing three short straws in a row, one trial consists of three draws. If the problem asks about flipping a coin that lands on heads four times in a row, one trial is four flips of the coin. This convention makes the computation for the probability more obvious.

Here's part of the introduction from lesson 103:

- The next box shows 2 hats with straws in them.
- (Teacher reference:)

◆ What's the probability of drawing a short straw from **both** of the hats?

◆ To figure out the probability of drawing a short straw from **both** hats, you **multiply.**

$$\dfrac{1}{3} \times \dfrac{1}{3} = \boxed{\dfrac{1}{9}}$$

- In each hat there are 3 straws: 1 short, 1 medium, 1 long.
- The problem asks: What's the probability of drawing a short straw from **both** of the hats?
- For this problem, you don't do just one thing for the trial. You draw a straw from the first hat **and** draw a straw from the second hat.

g. The fraction for the first hat is 1/3.
- What's the fraction for the second hat? (Signal.) *1/3.*
- To figure out the probability of drawing a short straw from both hats, you **multiply** the fractions.
- What do you do to figure out the probability of drawing a short straw from both hats? (Signal.) *Multiply the fractions.*
- (Repeat step g until firm.)

h. Let's say that there was a 3rd hat just like the first 2 and for each trial you had to draw a short straw from these **3** hats.
- Say the fraction for pulling a short straw from 1 hat. (Signal.) *1/3.*
- Say the multiplication for pulling a short straw from 2 hats. (Signal.) *1/3 times 1/3.*
- Say the multiplication for pulling a short straw from 3 hats. (Signal.) *1/3 times 1/3 times 1/3.*
- Raise your hand when you know what 1/3 times 1/3 times 1/3 equals. √
- What does it equal? (Signal.) *1/27.*
So the probability of someone drawing a short straw from all 3 hats is 1/27. It would only happen on about 1 of every 27 trials.

> **Teaching note:** In lesson 110, after students are firm in the problem-solving conventions, a teaching box shows why multiplication of individual probabilities works. For the probability of 1/3 × 1/3, a diagram shows that there are nine possibilities, but only one for the expected outcome.

Problems involving one event or another are introduced in lesson 105.

b. You've worked probability problems in which more than 1 thing happens on a trial. For those problems, you **multiply** the probabilities of each event.
- Some problems ask about the probability of 1 thing **or** another thing happening. For these problems, you **add** the probabilities of each event.

c. Remember, to find the probability of 1 event **and** another, you multiply.
- To find the probability of 1 event **or** another, you add.
- What do you do to find the probability of 1 event **and** another? (Signal.) *Multiply.*
- What do you do to find the probability of 1 event **or** another? (Signal.) *Add.*

d. You've already worked some probability problems that ask about 1 event or another. Some ask about the probability of a spinner landing on 1 color or another. Some ask about the probability of 1 side of a die or another.

---

A blindfolded person circles a day on a calendar that shows 1 week. What's the probability of the day being Tuesday **or** Saturday?

| S | M | T | W | T | F | S |

$\frac{1}{7} + \frac{1}{7} = \boxed{\frac{2}{7}}$

The chances of pulling a red object from a bag are 2 out of 8. The chances of pulling a blue object from the bag are 1 out of 8. What's the probability of pulling out a red object **or** a blue object?

---

e. The box shows a problem: A blindfolded person circles a day on a calendar that shows 1 week. What's the probability of the day being Tuesday or Saturday?
- The probability of a day being Tuesday is 1 out of 7. The probability of a day being Saturday is 1 out of 7. So the probability of a day being either Tuesday or Saturday is 1/7 + 1/7. That's 2/7.

f. The next problem tells about pulling objects from a bag: The chances of pulling a red object from a bag are 2 out of 8. The chances of pulling a blue object from the bag are 1 out of 8.
- Figure out the probability of pulling out a red object **or** a blue object. Raise your hand when you know the answer. √
- Everybody, what's the probability of pulling out a red object or a blue object? (Signal.) *3/8.*
Yes, 2/8 + 1/8 = 3/8.

The last problem type presented in this track begins in lesson 107. It involves changing the composition of a set on a sequence of trials. Students learn the rule that if you change the size of the group, you change the probabilities.

◆ If you change the size of a group, you change the probabilities.

◆ What's the probability of drawing a black ball?　○ ○ ● ○ ○ ○ ○

◆ What's the new probability?　○ ○ ● ○ ○ ○

◆ What's the new probability?　○ ○ ● ○

◆ What's the probability of drawing a red ball from this basket?

◆ What's the probability of drawing a white ball?

◆ Let's say a person pulls a white ball from the basket.

◆ What's the **new** probability of drawing a white ball?　◆ What's the **new** probability of drawing a red ball?

◆ When the total changes, the probability for each object in the group changes.

b. The box shows a rule for probabilities: If you change the size of a group, you change the probabilities.
- Say that rule. (Signal.) *If you change the size of a group, you change the probabilities.*
c. The top row shows 7 balls. One is black. So what's the probability of drawing a black ball from this group? (Signal.) *1/7.* Yes, 1 seventh.
- If we take 1 white ball from the group, what's the new group size? (Signal.) *6.*
- So what's the new probability of drawing a black ball from the group? (Signal.) *1/6.* The probability increases.

- If we remove 2 more white balls from the group, what's the new group size? (Signal.) *4.*
- What's the probability of drawing a black ball? (Signal.) *1/4.*
d. You can see a problem. There are 5 white balls and 3 red balls in a basket.
- Raise your hand when you know the probability of a person drawing a red ball from the basket. √
- Everybody, what's the probability? (Signal.) *3/8.* Yes, 3 eighths.
- What's the probability of drawing a white ball? (Signal.) *5/8.*
e. Let's say a person draws a white ball from the basket. Both probabilities have changed because there is now a different number for the whole group and a different number for the white balls.
- Raise your hand when you know the new probability of drawing a white ball. √
- What's the probability of drawing a white ball? (Signal.) *4/7.*
- Raise your hand when you know the new probability for drawing a red ball. √
- What's the probability of drawing a red ball? (Signal.) *3/7.*
f. Remember, when the total changes, the probability for each object in the group changes.

Word problems begin in lesson 108. Here are two of the problems students work in 108:

A person draws a medium sweatshirt from the dryer but does not return it. What's the probability that the next person will pull out a sweatshirt that is large?

A person draws all 3 black sweatshirts from the dryer and does not return them. What's the probability that the next person will pull out a white sweatshirt?

In lesson 109, students work problems that involve two events and the composition of the group changing. Here's part of the introduction from lesson 109:

b. The box shows a new kind of problem: There are 3 apples and 7 plums on a shelf. A girl randomly picks a fruit from the shelf and eats it. Then the girl randomly picks a second fruit from the shelf. What is the probability of the girl first eating an apple and then picking a plum?
• This problem tells about doing 2 things in a row, so you multiply. But the number of things on the shelf changes because the girl eats one of them.
c. Raise your hand when you know the probability of the girl eating an apple first. √
• Everybody, what's the probability? (Signal.) *3/10.*
d. The size of the group has changed. Raise your hand when you know the new probability for eating a plum. √
• What's the new probability for eating a plum? (Signal.) *7/9.*
• So we multiply 3/10 by 7/9.
e. Figure out the probability of those two events occurring one after the other. √
• (Write on the board:)                    [109:5A]

$$\frac{\cancel{3}}{10} \times \frac{7}{\cancel{9}_3} = \boxed{\frac{7}{30}}$$

• Here's what you should have. You multiply 3/10 times 7/9. The simplified answer is 7/30.

## Scientific Notation (Lessons 89–112)

In lessons 89–99, students work with large numbers.

They write numbers for the scientific notation. For $5.3 \times 10^4$, they write 53,000. Starting in lesson 92, students write the scientific notation for numbers.

Starting in lesson 104, students work with small numbers, such as .0000454. Students write the scientific notation ($4.54 \times 10^{-5}$). Starting with lesson 107, students write numbers for the scientific notation. For the final applications (109–112), students rank a set of values shown in scientific notation from largest to smallest. The set includes values with positive exponents and values with negative exponents.

The first skill students learn is to write large numbers from scientific notation. They learn that the exponent tells the number of places to move the decimal point. Here's part of the introduction from lesson 89:

---

◆ When you multiply any value by a power of 10, you can figure out the answer by moving the decimal point of that value.

◆ The exponent of 10 tells you how many places to move the decimal point.

---

$2.3 \times 10^4 = 2.3000$    ◆ Move the decimal point 4 places.

$= \boxed{23,000}$

---

b. I'll read what it says. Follow along: When you multiply any value by a power of 10, you can figure out the answer by moving the decimal point of that value.
• Once more: When you multiply any value by a power of 10, you can figure out the answer by moving the decimal point of that value.
c. The box shows 2.3 times $10^4$.
• The exponent of $10^4$ tells you to move the decimal point 4 places. How many places do you move the decimal point if you multiply by $10^4$? (Signal.) *4.*
• How many places would you move the decimal point for $10^3$? (Signal.) *3.*
• How many places would you move the decimal point for $10^5$? (Signal.) *5.*
d. You can see the number you get when you multiply 2.3 by $10^4$. The decimal point of 2.3 is moved 4 places. So the answer is 23,000.
• 2.3 times $10^4$ equals 23,000.
• Read that equation. (Signal.) *2.3 times $10^4$ equals 23,000.*

Students copy problems written in scientific notation and write the number for each notation.

In lesson 89, students also figure out a missing exponent of 10:

$$3.45 \times 10^{\blacksquare} = 345{,}000$$
$$1.028 \times 10^{\blacksquare} = 10{,}280$$

Students count from the place after the first digit to after the last digit.

In lesson 92, students write the scientific notation for numbers. Items are presented in this form:

$$75{,}200 = \blacksquare \times 10^{\blacksquare}.$$

The rules students follow are these: Copy the digits before the zeros; write a decimal point after the first digit; write the exponent for 10.

Here's part of the introduction:

> a. $752{,}000 = \boxed{\phantom{xxx}} \times 10^{\blacksquare}$
>
> b. $7{,}131{,}000 = \boxed{\phantom{xxx}} \times 10^{\blacksquare}$

b. Problem A: 752,000.
- What are the digits before the zeros? (Signal.) *7, 5, 2.*
- So what decimal value do you write in the first box? (Signal.) *7.52.*

g. After you write the decimal value in the first box, you write the exponent for 10.
- You just start after the first digit of the original value and write the number of places to the end of the original value.

h. (Write on the board:)  [92:1A]

> a. $752{,}000 = \boxed{\phantom{xxx}} \times 10^{\square}$

- Here's problem A. Tell me the decimal value that goes in the first box. (Signal.) *7.52.*

- (Write to show:)  [92:1B]

> a. $752{,}000 = \boxed{7.52} \times 10^{\square}$

- Now I start after the first digit of 752,000 and count the places to the end of the number.
- Raise your hand when you know the number of places. √
- Everybody, how many places? (Signal.) *5.*
- So the exponent for 10 is 5.
- (Write to show:)  [92:1C]

> a. $752{,}000 = \boxed{7.52} \times 10^{\boxed{5}}$

- 752,000 equals 7.52 times $10^5$.
i. Your turn: Complete item A. Then do item B. Pencils down when you're finished. √

In subsequent lessons, students learn the expression *scientific notation* and work with values that have a decimal point and values that do not end in zero. Students copy all the digits.

Conventions for values less than one are presented in lesson 104. Here's part of the introduction:

> ◆ Copy the digits that come after the zeros.     $.000064 = \blacksquare \times 10^{\blacksquare}$
>
> ◆ Write the decimal point after the first digit.     $= 6.4 \times 10^{\blacksquare}$
>
> ◆ If the original number is **less than 1,** the exponent is **negative.**     $= 6.4 \times 10^{-5}$

b. You can write very small numbers in scientific notation. The numbers we'll work with are greater than zero, but less than 1.
- The number in the box is .000064. This value is less than 1, so we copy the digits that come **after the zeros.** Those digits are 6 and 4. Then we put the decimal point after the first digit. We now have 6.4 times 10 to some power.

c. The rule in the box tells about the exponent of 10: If the original number is **less than 1,** the exponent is **negative.**
- Once more. If the original number is less than 1, what do you know about the exponent? (Signal.) *It's negative.*
- We started with a number that is less than 1, so we just count the number of places from the new decimal point and write the negative exponent.
- Raise your hand when you know how many places we moved the decimal point. √
- How many places? (Signal.) *5.*
  So the exponent is **minus** 5.
d. Remember, to write scientific notation for a value that is less than 1, you write a negative exponent, not a positive one.

In lesson 107, students write numbers that are less than one from scientific notation. The procedure students use is the opposite of that for writing large numbers from scientific notation. Students copy the digits, start after the first digit, and count places to the **left.**

b. You've written scientific notation for small numbers less than 1.
- For those numbers, the exponent of 10 is negative.
- When you write a number from scientific notation that has a negative exponent, you copy the digits shown for the scientific notation. Then start after the first digit and count places to the **left.**
c. The box shows the scientific notation $3.08 \times 10^{-5}$.
- (Teacher reference:)

$$308 = 3.08 \times 10^{-5}$$
$$308 =$$
$$.000308 =$$

- You can see the digits 3-0-8, and five places to the left of the first digit.

- You write the new decimal point and a zero for each empty space. There are 4 empty spaces, so you write 4 zeros.
- Remember, copy the digits in the scientific notation. Then start after the first digit and count places to the left.

| a. | 712 | $= 7.12 \times 10^{-6}$ |

d. Problem A: The scientific notation is $7.12 \times 10^{-6}$.
- The digits 7, 1, 2 are already shown on the left. Start after 7 and count to the left. The exponent of 10 is minus 6, so how many places do you count to the left? (Signal.) *6.*
- Count the places. Make a new decimal point. Then write a zero for each empty space. Pencils down when you're finished. (Observe students and give feedback.)
- (Write on the board:) [107:4A]

| a. | .00000712 |

- Here's the number you should have.
- How many zeros did you make? (Signal.) *5.*

In lessons 109 and 110, students rank values that are more than one and less than one. Values are expressed in scientific notation.

Here's part of the exercise from lesson 109:

| _____ $5.72 \times 10^{-12}$ | _____ $6.78 \times 10^{17}$ |
| _____ $8.524 \times 10^{12}$ | _____ $2.16 \times 10^{-11}$ |
| _____ $4.85 \times 10^{-12}$ | _____ $6.52 \times 10^{17}$ |
| _____ $7.4 \times 10^{15}$ | _____ $6.69 \times 10^{-9}$ |

- All these values are expressed in scientific notation. Some are very big numbers and some are very small.

b. Your turn: Find the largest number. First look for the value with the largest exponent. Then make sure that the decimal number is bigger than any other that has the same power of 10. Raise your hand when you've found the largest value. √
- Everybody, which value is the largest? (Signal.) *6.78 × 10¹⁷.*

c. There's another value that is almost as large. Touch that value. √
- Everybody, what's the second largest value? (Signal.) *6.52 × 10¹⁷.*

d. Now you're going to find the smallest value. √
- Find the smallest value. That's the value that has the negative exponent farthest from zero. If there are 2 values with the same exponent, find the decimal value with the smaller number. Raise your hand when you've found the smallest value. √
- Everybody, which number has the smallest value? (Signal.) *4.85 × 10⁻¹².*

e. There's another value that is almost as small as that. Touch that value. √
- Everybody, what's the second smallest value? (Signal.) *5.72 × 10⁻¹².*
- (Repeat steps b–e until firm.)

f. Write 1 by the largest value, 2 by the next largest value, and keep on going through 8. Pencils down when you're finished. (Observe students and give feedback.)

**Key:**

$$
\begin{array}{cl}
\underline{7} & 5.72 \times 10^{-12} \\
\underline{4} & 8.524 \times 10^{12} \\
\underline{8} & 4.85 \times 10^{-12} \\
\underline{3} & 7.4 \times 10^{15} \\
\underline{1} & 6.78 \times 10^{17} \\
\underline{6} & 2.16 \times 10^{-11} \\
\underline{2} & 6.52 \times 10^{17} \\
\underline{5} & 6.69 \times 10^{-9}
\end{array}
$$

For the final task presented in lesson 112, students complete a list so that all values are shown in scientific notation, then rank the values from largest to smallest. Here's the set of values given in lesson 112:

| | |
|---|---|
| **a.** $9 \times 10^6$ | **d.** .4203 |
| **b.** $4.36 \times 10^{-3}$ | **e.** $1.12 \times 10^1$ |
| **c.** 50,000,000 | |

# Proportion (Lessons 100–103)

This track is a simple extension of what students have learned about similar triangles. The extension is to similar figures. Just as corresponding sides of similar triangles are proportional, corresponding dimensions of similar figures are proportional.

For the first set of problems, students are given three dimensions for a pair of similar figures; they figure out the fourth. For the later problems, students work from scale diagrams of objects. They are given a fact about the real-life dimensions of one object. They compute the corresponding dimension of a comparison object.

Here's the introduction to proportion from lesson 100:

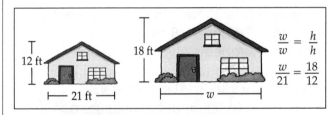

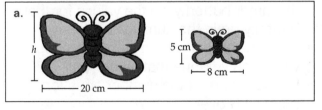

b. These are called **proportion problems.**
- What are they called? (Signal.) *Proportion problems.*
- You work them the same way you work similar triangle problems.

c. The problem in the box shows houses that are similar. They are the same shape but different sizes. The smaller house is 21 feet wide and 12 feet high.
- One measurement is given for the larger house. Is it the width or the height? (Signal.) *Height.*
  Yes, the height.
- You have to figure out the width of the larger house. So the values for the larger house are in the numerators.
- You can see the equation W/W = H/H. Below is the equation with numbers. The first fraction shows the corresponding widths: W over 21.
- The other fraction shows the corresponding heights: 18/12.

d. Copy the equation and figure out the width of the larger house. Pencils down when you're finished.
  (Observe students and give feedback.)
- Check your work.
- What's the width of the larger house? (Signal.) *31 and 1/2 feet.*

e. Problem A is like the one in the box. The height of the little butterfly is 5 centimeters. What's the width of the little butterfly? (Signal.) *8 centimeters.*
- Which number is given for the big butterfly, height or width? (Signal.) *Width.*
  Yes, the width is 20 centimeters.
- You have to figure out a measurement for the larger butterfly, so the values for that butterfly go in the numerator.
- Write the equation with numbers. Stop when you've done that much. √
- Everybody, read the number equation. (Signal.) *H over 5 = 20/8.*
- (Write on the board:)          [100:4A]

$$\frac{h}{5} = \frac{20}{8}$$

- The number equation is H over 5 = 20/8.
- Solve the equation and write the height of the larger butterfly. Pencils down when you're finished.
  (Observe students and give feedback.)

- Check your work.
- What's the height of the larger butterfly? (Signal.) *12 and 1/2 centimeters.*

In lesson 102, students work with dimensions of real-life objects. Here's part of the introduction from lesson 102:

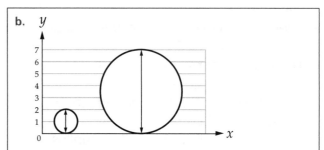

**b.**

**Fact:** The smaller sphere has a diameter of 36 centimeters in real life.

**Question:** What's the actual diameter of the larger sphere?

d. Item B. The diagram shows 2 spheres. There are 2 similar spheres in real life. The fact gives the diameter of the smaller sphere in real life. You have to figure out the **actual** diameter of the larger sphere.
- Touch the diameter of the smaller sphere. √
  How many units is that diameter? (Signal.) *2.*
- Touch the diameter of the larger sphere. √
  How many units is that diameter? (Signal.) *7.*
- The fact says: The smaller sphere has a diameter of 36 centimeters in real life.
- You'll figure out the actual diameter of the larger sphere. To work the problem, you write 1 fraction that shows the corresponding diameters of the actual spheres and a fraction for the corresponding diameters of the spheres on the grid.
- Write the equation and figure out the actual diameter of the larger sphere. Pencils down when you're finished.
  (Observe students and give feedback.)

- (Write on the board:)  [102:2C]

**b.** $(36)\dfrac{d}{36} = \dfrac{7}{2}(36)$

$d = 126$

| 126 centimeters |

- Here's what you should have.
- Everybody, what's the actual diameter of the larger sphere? (Signal.) *126 centimeters.*

**Teaching note:** Students apply what they have learned about the coordinate system to figure out the number of units for the diameter of each circle. Students are required to specify corresponding dimensions of two pairs of similar objects (two diagrammed objects and two real-life objects). If students get confused, tell them to label one fraction *diagram* and the other *real-life.* In both fractions, the numerator shows the dimensions for the larger sphere.

# Box and Whiskers (Lessons 111–118)

Students learn to describe a population using a box between the upper and lower quartiles and lines (whiskers) to show the range of the population. The track teaches mean, median, and definitions of the quartiles.

The track starts with mean and median. To find the mean, students add all the scores and divide by the number of scores. To find the median score, students use the total number of scores in the population to locate the position of the middle score.

Here's part of the introduction:

| ◆ The **median** is the **middle** score. |
| --- |
| 2, 5, 6, 7, 11 |
| 12, 11, 10, 10, 9, 4, 3, 1 |
| ◆ If there is an **even number of scores,** the middle score is not shown. It is **midway between two scores**. |

b. Another measure of populations is the median.
- **Median** means **middle.** The median score is the middle score.
- What's the middle score called? (Signal.) *The median score.*

c. Look at the first set of scores.
- These scores are arranged from lowest to highest.
- How many scores are in that group? (Signal.) *5.*
- One of those scores is the median score. It's in the middle. Raise your hand when you know the median score. √
- Everybody, what's the median score? (Signal.) *6.*
  Yes, 6 is the middle score. It has two scores on either side.
d. Look at the next set of scores.
- They are arranged from highest to lowest.
- Raise your hand when you know how many scores are in that set. √
- How many scores are in that set? (Signal.) *8.*
- 8 is an even number. So the middle score is not shown. It is between two scores.
- Find the point where you have the same number of scores on each side. √
- Everybody, the median score is between which two numbers? (Signal.) *10 and 9.*
  Yes, the median score is midway between 10 and 9. So the median score is 9.5.
- What's the median score? (Signal.) *9.5.*
- How many scores are on each side of the median score? (Signal.) *4.*
e. Remember, the median score has the same number of scores on either side. If the score is between 4 and 5, the median score is 4.5.
- If the median score is between 10 and 12, it is midway between those numbers. So what's the median score? (Signal.) *11.*
- If the median score is between 8 and 8, what's the median score? (Signal.) *8.*
- If the median score is between 20 and 21, what's the median score? (Signal.) *20.5.*
- If the median score is between 20 and 30, what's the median score? (Signal.) *25.*
- (Repeat step e until firm.)
f. Remember, to find the median, you first make sure the scores are ordered from lowest to highest or highest to lowest. Then you figure out the middle score.
- If there is an **even number of scores,** the middle score is not shown. It is **midway between two scores.**

In following lessons, students find the mean and the median of populations. Quartiles are introduced in lesson 115. The lower quartile is the median for the lower half of the population. The upper quartile is the median for the upper half of the population. Here's part of the introduction that indicates the percentage of scores above and below the quartiles:

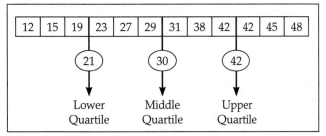

d. Touch the line for the lower quartile again. √
   • 1/4 of the scores is below the lower quartile. That's 25% of the scores.
   • What percent of the scores is below the lower quartile? (Signal.) *25%.*
   • What percent of the scores is above that score? (Signal.) *75%.*
   • (Repeat step d until firm.)
e. Touch the middle quartile. √
   • What percent of the scores is below the middle quartile? (Signal.) *50%.*
   • What percent of the scores is above that score? (Signal.) *50%.*
f. Touch the upper quartile. √
   • What percent of the scores is below the upper quartile? (Signal.) *75%.*
   • What percent of the scores is above that score? (Signal.) *25%.*
   • (Repeat steps e and f until firm.)
g. Remember, the quartile scores are the median scores that divide a population into 4 parts.

Lesson 117 introduces the box and whiskers diagram.

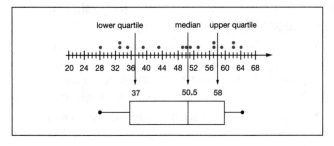

b. The box shows a way of displaying a population of scores. It's called box and whiskers.
   • What's it called? (Signal.) *Box and whiskers.*
   • Touch the box at the bottom. √
   • That box extends from the lower quartile to the upper quartile of the population.
   • How far does the box extend? (Signal.) *From the lower quartile to the upper quartile.*
c. Touch the end of the box at the lower quartile of the population. √
   • What's the score for that end of the box? (Signal.) *37.*
d. Touch the end of the box at the upper quartile of the population. √
   • What's the score for that end of the box? (Signal.) *58.*
e. Remember, the box extends from the lower quartile to the upper quartile of the population.
f. The box and whiskers also shows the median score. It's the line inside the box. Touch the line for the median score. √
   • What's the median score of the population? (Signal.) *50.5.*

Students use the following steps when constructing a box and whiskers for a population:

   • They mark the median, the lower quartile, and the upper quartile.

   • They make a point for the highest score and the lowest score.

   • They draw a box that extends from the lower quartile to the upper quartile.

   • They make a line inside the box for the median score.

   • Finally, they draw a line from each side of the box to the lowest score and the highest score.

# Test Preparation: Lessons 119 and 120

## Lessons 119 and 120

The second half of *Essentials for Algebra* focuses increasingly on problem types that appear on exit exams. The last two lessons of the program (lessons 119 and 120) present items in the multiple-choice format that appears on exit exams and achievement tests. The lessons in which this test preparation occurs have a different structure from preceding lessons. Students first work a series of items that are loosely related to a central theme, such as data analysis or geometry. Some of the items are like those students have worked previously. Others involve extensions or variations of familiar material.

Here's part of lesson 119 that involves scatterplots.

a. Open your textbook to lesson 119, part 1. √

• These are like items you'll see on tests. The items have some language that is new. I'll show you the answers to items or how to work items. Then you'll work other items of the same type.

• For some of these items, you'll use your ruler.

b. Item 1: Of the following scatterplots, which shows a negative correlation?

• Each graph works like a coordinate system. It shows a population of points. A positive correlation is a positive slope. A negative correlation is a negative slope.

c. Raise your hand when you know which graph shows a **negative** correlation. √

• Everybody, which graph? (Signal.) *B.*

d. Raise your hand when you know which 2 graphs show a positive correlation. √

• Everybody, which graphs show a positive correlation? (Signal.) *A and D.*

e. Graph C doesn't have a positive correlation or a negative correlation.

f. Find Item 2. √

g. Item 2: The cost of a bushel of corn varies according to the number of bushels purchased. Which of the following conclusions about the number of bushels purchased and the cost per bushel is best supported by the scatterplot above?

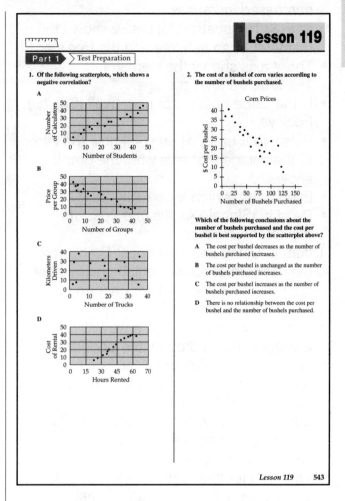

• The Y axis is the cost per bushel.
• What's the X axis? (Signal.) *Number of bushels purchased.*
• Point to show the direction on the graph for the cost of bushels increasing. (Students point up.) √
• Point to show the direction for the cost of bushels decreasing. (Students point down.) √
• Raise your hand when you know if the slope of points on this graph is positive or negative. √
• What's the slope? (Signal.) *Negative.* So the cost per bushel **decreases** as the number of bushels purchased increases.
• Raise your hand when you know the letter of the choice that tells what the graph shows. √
• Which letter? (Signal.) *A.*

- Read it. (Signal.) *The cost per bushel decreases as the number of bushels purchased increases.*
h. Item 3: The scatterplot below shows the diameter of an ash tree and the tree's age.
- The Y axis shows the diameter. The X axis shows the age.
- Is the slope positive or negative? (Signal.) *Positive.*
- Raise your hand when you know the letter of the choice that tells what the graph shows. √
- Everybody, which letter? (Signal.) *C.*
- Read it. (Signal.) *As the tree ages, its diameter increases.*
i. Item 4: Using the line of best fit shown on the scatterplot above, which choice best approximates the cost per uniform for a team with 40 players?
- The line of best fit is just a way of averaging the scores for the population.
- The question asks about the rate per uniform for 40 players. To find the rate, you go to 40 on the X axis.
- Use your ruler to go up to the line. Then go across to find the corresponding Y value. You write the rate as Y over X.
- Write the fraction based on the graph. √
- What does Y over X equal? (Signal.) *600/40.*
- Now simplify the fraction. √
- Find the letter of the choice that gives the right cost per uniform.
- Everybody, which letter? (Signal.) *D.*
- What's the answer? (Signal.) *$15.*
j. Item 5: The graph below shows the price for a steer from the Lilly Ranch at the end of every other year from 1999 to 2005.
- What's the Y axis? (Signal.) *Cost.* Yes, cost in dollars.
- What's the X axis? (Signal.) *Years.*
- The item asks: From this graph, which of the following was the most probable price of a steer from the Lilly Ranch at the end of 1997?

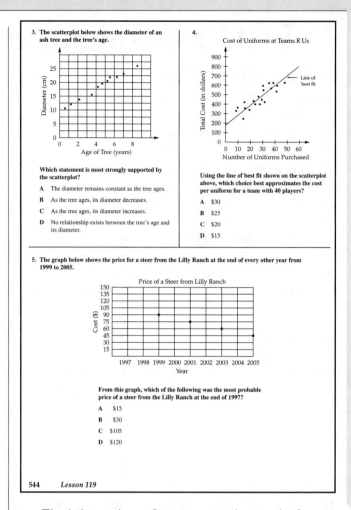

3. The scatterplot below shows the diameter of an ash tree and the tree's age.

Diameter (cm) / Age of Tree (years)

Which statement is most strongly supported by the scatterplot?

A   The diameter remains constant as the tree ages.

B   As the tree ages, its diameter decreases.

C   As the tree ages, its diameter increases.

D   No relationship exists between the tree's age and its diameter.

4. Cost of Uniforms at Teams R Us

Total Cost (in dollars) / Number of Uniforms Purchased — Line of best fit

Using the line of best fit shown on the scatterplot above, which choice best approximates the cost per uniform for a team with 40 players?

A   $30

B   $25

C   $20

D   $15

5. The graph below shows the price for a steer from the Lilly Ranch at the end of every other year from 1999 to 2005.

Price of a Steer from Lilly Ranch

Cost ($) / Year

From this graph, which of the following was the most probable price of a steer from the Lilly Ranch at the end of 1997?

A   $15

B   $30

C   $105

D   $120

544     *Lesson 119*

- Find the value of a steer at the end of 1999. √
- Everybody, what's that value? (Signal.) *$90.*
- Find the probable value of a steer at the end of 1997. There's no point there. You can use your ruler to show the line. Raise your hand when you know the letter of the right choice. √
- Everybody, which letter? (Signal.) *C.*
- What's the probable price of a steer at the end of 1997? (Signal.) *$105.*

**Teaching note:** Item 1: Although students have not previously worked with scatterplots or correlations, they understand slope. Therefore, the explanation of positive and negative correlations is straightforward and not difficult for students to understand. The positive slope shows a positive correlation.

Item 2: The application is new, but the fundamental relations are familiar. For a positive slope, $y$ increases as $x$ increases; for a negative slope, $y$ decreases as $x$ increases. Item 2 presents a negative correlation, which is described as cost decreasing as the number of bushels purchased increases.

Item 4: This problem presents a new rate application. The line of best fit is shown and may be used as a basis for the rate. The procedure for determining the rate is familiar. Students write the fraction $y$ over $x$ for the values identified by the question. Students then simplify. Although this is a new application, students have the math skills necessary to solve the problem.

Item 5: The item asks about a probable value of a steer at a point in time. To work the problem, students simply extrapolate on the basis of the pattern shown. The procedure involves extending a line to the particular point in time.

Although the format for responding to the items requires specifying one of the choices, students work each problem that involves calculation.

For this exercise, after you have presented the six problems in part 1, students work parallel problems in part 2 (items 7–13) that call for bubblesheet responses. For this work, you provide no structure. Students work all seven problems and then check all seven problems.

Expect some students to have trouble with some of the problems. Repeat any problems that present serious difficulties.

## Practice Tests

Following lesson 120, students will receive two practice tests that have a full range of the problem types presented in lessons 119 and 120, plus some additional problems students should be able to solve based on what they have learned. Allow students two 55-minute class periods for each test. Possibly the best way to give feedback is to collect the students' work for the test and mark mistakes in the student workbooks. Note items that are missed by 25% or more of the students. Go over these items. Students who missed one quarter or more of the items should then repeat the entire test.

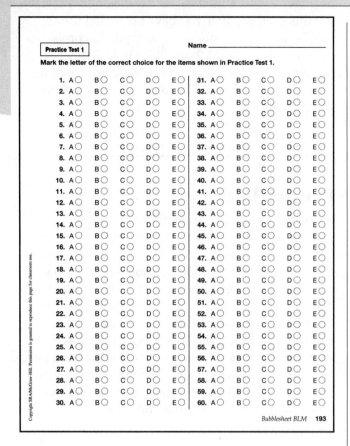

**Practice Test 1**

Name _____

Mark the letter of the correct choice for the items shown in Practice Test 1.

| | | | | | | | | | |
|---|---|---|---|---|---|---|---|---|---|
| 1. A○ | B○ | C○ | D○ | E○ | 31. A○ | B○ | C○ | D○ | E○ |
| 2. A○ | B○ | C○ | D○ | E○ | 32. A○ | B○ | C○ | D○ | E○ |
| 3. A○ | B○ | C○ | D○ | E○ | 33. A○ | B○ | C○ | D○ | E○ |
| 4. A○ | B○ | C○ | D○ | E○ | 34. A○ | B○ | C○ | D○ | E○ |
| 5. A○ | B○ | C○ | D○ | E○ | 35. A○ | B○ | C○ | D○ | E○ |
| 6. A○ | B○ | C○ | D○ | E○ | 36. A○ | B○ | C○ | D○ | E○ |
| 7. A○ | B○ | C○ | D○ | E○ | 37. A○ | B○ | C○ | D○ | E○ |
| 8. A○ | B○ | C○ | D○ | E○ | 38. A○ | B○ | C○ | D○ | E○ |
| 9. A○ | B○ | C○ | D○ | E○ | 39. A○ | B○ | C○ | D○ | E○ |
| 10. A○ | B○ | C○ | D○ | E○ | 40. A○ | B○ | C○ | D○ | E○ |
| 11. A○ | B○ | C○ | D○ | E○ | 41. A○ | B○ | C○ | D○ | E○ |
| 12. A○ | B○ | C○ | D○ | E○ | 42. A○ | B○ | C○ | D○ | E○ |
| 13. A○ | B○ | C○ | D○ | E○ | 43. A○ | B○ | C○ | D○ | E○ |
| 14. A○ | B○ | C○ | D○ | E○ | 44. A○ | B○ | C○ | D○ | E○ |
| 15. A○ | B○ | C○ | D○ | E○ | 45. A○ | B○ | C○ | D○ | E○ |
| 16. A○ | B○ | C○ | D○ | E○ | 46. A○ | B○ | C○ | D○ | E○ |
| 17. A○ | B○ | C○ | D○ | E○ | 47. A○ | B○ | C○ | D○ | E○ |
| 18. A○ | B○ | C○ | D○ | E○ | 48. A○ | B○ | C○ | D○ | E○ |
| 19. A○ | B○ | C○ | D○ | E○ | 49. A○ | B○ | C○ | D○ | E○ |
| 20. A○ | B○ | C○ | D○ | E○ | 50. A○ | B○ | C○ | D○ | E○ |
| 21. A○ | B○ | C○ | D○ | E○ | 51. A○ | B○ | C○ | D○ | E○ |
| 22. A○ | B○ | C○ | D○ | E○ | 52. A○ | B○ | C○ | D○ | E○ |
| 23. A○ | B○ | C○ | D○ | E○ | 53. A○ | B○ | C○ | D○ | E○ |
| 24. A○ | B○ | C○ | D○ | E○ | 54. A○ | B○ | C○ | D○ | E○ |
| 25. A○ | B○ | C○ | D○ | E○ | 55. A○ | B○ | C○ | D○ | E○ |
| 26. A○ | B○ | C○ | D○ | E○ | 56. A○ | B○ | C○ | D○ | E○ |
| 27. A○ | B○ | C○ | D○ | E○ | 57. A○ | B○ | C○ | D○ | E○ |
| 28. A○ | B○ | C○ | D○ | E○ | 58. A○ | B○ | C○ | D○ | E○ |
| 29. A○ | B○ | C○ | D○ | E○ | 59. A○ | B○ | C○ | D○ | E○ |
| 30. A○ | B○ | C○ | D○ | E○ | 60. A○ | B○ | C○ | D○ | E○ |

*Bubblesheet BLM* **193**

**Practice Test 2**

Name _____

Mark the letter of the correct choice for the items shown in Practice Test 2.

| | | | | | | | | | |
|---|---|---|---|---|---|---|---|---|---|
| 1. A○ | B○ | C○ | D○ | E○ | 31. A○ | B○ | C○ | D○ | E○ |
| 2. A○ | B○ | C○ | D○ | E○ | 32. A○ | B○ | C○ | D○ | E○ |
| 3. A○ | B○ | C○ | D○ | E○ | 33. A○ | B○ | C○ | D○ | E○ |
| 4. A○ | B○ | C○ | D○ | E○ | 34. A○ | B○ | C○ | D○ | E○ |
| 5. A○ | B○ | C○ | D○ | E○ | 35. A○ | B○ | C○ | D○ | E○ |
| 6. A○ | B○ | C○ | D○ | E○ | 36. A○ | B○ | C○ | D○ | E○ |
| 7. A○ | B○ | C○ | D○ | E○ | 37. A○ | B○ | C○ | D○ | E○ |
| 8. A○ | B○ | C○ | D○ | E○ | 38. A○ | B○ | C○ | D○ | E○ |
| 9. A○ | B○ | C○ | D○ | E○ | 39. A○ | B○ | C○ | D○ | E○ |
| 10. A○ | B○ | C○ | D○ | E○ | 40. A○ | B○ | C○ | D○ | E○ |
| 11. A○ | B○ | C○ | D○ | E○ | 41. A○ | B○ | C○ | D○ | E○ |
| 12. A○ | B○ | C○ | D○ | E○ | 42. A○ | B○ | C○ | D○ | E○ |
| 13. A○ | B○ | C○ | D○ | E○ | 43. A○ | B○ | C○ | D○ | E○ |
| 14. A○ | B○ | C○ | D○ | E○ | 44. A○ | B○ | C○ | D○ | E○ |
| 15. A○ | B○ | C○ | D○ | E○ | 45. A○ | B○ | C○ | D○ | E○ |
| 16. A○ | B○ | C○ | D○ | E○ | 46. A○ | B○ | C○ | D○ | E○ |
| 17. A○ | B○ | C○ | D○ | E○ | 47. A○ | B○ | C○ | D○ | E○ |
| 18. A○ | B○ | C○ | D○ | E○ | 48. A○ | B○ | C○ | D○ | E○ |
| 19. A○ | B○ | C○ | D○ | E○ | 49. A○ | B○ | C○ | D○ | E○ |
| 20. A○ | B○ | C○ | D○ | E○ | 50. A○ | B○ | C○ | D○ | E○ |
| 21. A○ | B○ | C○ | D○ | E○ | 51. A○ | B○ | C○ | D○ | E○ |
| 22. A○ | B○ | C○ | D○ | E○ | 52. A○ | B○ | C○ | D○ | E○ |
| 23. A○ | B○ | C○ | D○ | E○ | 53. A○ | B○ | C○ | D○ | E○ |
| 24. A○ | B○ | C○ | D○ | E○ | 54. A○ | B○ | C○ | D○ | E○ |
| 25. A○ | B○ | C○ | D○ | E○ | 55. A○ | B○ | C○ | D○ | E○ |
| 26. A○ | B○ | C○ | D○ | E○ | 56. A○ | B○ | C○ | D○ | E○ |
| 27. A○ | B○ | C○ | D○ | E○ | 57. A○ | B○ | C○ | D○ | E○ |
| 28. A○ | B○ | C○ | D○ | E○ | 58. A○ | B○ | C○ | D○ | E○ |
| 29. A○ | B○ | C○ | D○ | E○ | 59. A○ | B○ | C○ | D○ | E○ |
| 30. A○ | B○ | C○ | D○ | E○ | 60. A○ | B○ | C○ | D○ | E○ |

**194** *Bubblesheet BLM*

Two reproducible bubblesheets appear at the end of this Guide. Use these when students retake a test.

With this preparation, students should be able to apply what they have learned in *Essentials* to graduation math tests.

# Placement Test

The Placement Test provides for three outcomes:

1. The student lacks the necessary skills to place in *Essentials for Algebra.*

2. The student places at lesson 1 of *Essentials for Algebra.*

3. The student places at lesson 16 of *Essentials for Algebra.*

The test has two sections: A and B.

1. Results from Section A determine the following:

   a) The student lacks the necessary skills to place in *Essentials for Algebra.*

   b) The student places at lesson 1.

   c) The student should take Section B of the Placement Test.

2. Results from Section B determine the following:

   a) The student places at lesson 1.

   b) The student places at lesson 16.

## Administering the Placement Test

### DIRECTIONS

I'm going to pass out a test. Do not write anything until I tell you to start.

*(Pass out section A.)*

Write your name at the top.

You'll write your answers in pen, not pencil. You can use the front of the sheet or the back to figure out answers. Write your final answer to each problem in the box or where the answer belongs.

You have 12 minutes to complete the test, starting now.

*(Time students. Monitor their performance.)*

*(At the end of 12 minutes)*

Stop.

You're going to score your own tests. Exchange papers with a neighbor. √

You'll write **C** next to each answer that is correct and **X** next to each answer that is wrong.

Part 1, item **a**: The answer is 4.

*(Repeat for remaining items.)*

> ***Key:*** **b.** zero   **c.** 5   **d.** 3   **e.** 7   **f.** 2
> **g.** 127 should not be circled.
> **h.** 48 should be circled.
> **i.** 150 should be circled.
> **j.** 483 should not be circled.
> **k.** 536 should be circled.
> **l.** 5   **m.** 8   **n.** 6   **o.** 3   **p.** 7   **q.** 7   **r.** 3
> **s.** 8   **t.** 9   **u.** 6

Part 2, Item **a**: The answer is 750.

*(Repeat for remaining items.)*

> ***Key:*** **b:** 2798   **c:** 232   **d:** 364   **e:** 321   **f:** 379
> **g:** 302   **h:** 1/4   **i:** 6/5 or 1 and 1/5
> **j:** 5/72   **k:** 42/25 or 1 and 17/25.

Count all items marked with an **X.**

Write that number in the score box at the top of the test.

*(Observe students and give feedback.)*

*(Collect papers.)*

*(Pass out Section B to all students who missed 0–8 items.)*

Raise your hand when you're finished. I'll pick up your test.

*(Students who take Section B may require 30 minutes to complete all items.)*

*(Collect and grade tests.)*

# Placement Test Criteria

## SECTION A

Section A consists of 2 parts and 32 items. Placement is determined by totaling the errors made.

### CRITERIA

- Students who miss a total of 0–8 items should take Section B of the Placement Test.

- Students who miss a total of 9–11 items place at Lesson 1.

- Students who miss a total of 12 or more items do not have sufficient skills to place in *Essentials for Algebra.*

Section B consists of 8 parts and 40 items. All the items in Section B involve skills that are taught in the first 15 lessons of *Essentials for Algebra.* Students who demonstrate these skills are placed at Lesson 16. Students who do not demonstrate sufficient mastery are placed at Lesson 1. A student is placed at Lesson 1 by failing too many parts or by failing too many items.

### CRITERIA FOR PARTS

- Passing criterion for Parts 1–4: Students make no more than 2 errors on each part.

- Passing criterion for Part 5: Students make errors on no more than 2 **rows.**

- Passing criterion for Part 6: Students make errors on no more than 2 **equations.**

- Passing criterion for Parts 7 and 8: Students make no more than 1 error on each part.

### PLACEMENT CRITERIA

- Students who fail 3 or more parts place at Lesson 1.

- Students who make 12 or more errors place at Lesson 1.

- Students who fail 2 or fewer parts **and** who make 11 or fewer total errors place at Lesson 16.

Use the Placement Test Summary Chart on page 159 to record each student's performance and determine the placement for each student.

# Placement Test Summary Chart

| Student Names | | | | | | | | | | | |
|---|---|---|---|---|---|---|---|---|---|---|---|
| **Section A** (number of errors) | | | | | | | | | | | |
| Select 1 of these | 12 or more errors: **Insufficient skills** | | | | | | | | | | |
| | 9–11 errors: **Place at Lesson 1.** | | | | | | | | | | |
| | 0–8 errors: **Present Section B.** | | | | | | | | | | |
| **Section B** | | | | | | | | | | | |
| **Part 1** (6 items) number of errors Check if 3 or more errors. | | | | | | | | | | | |
| **Part 2** (6 items) number of errors Check if 3 or more errors. | | | | | | | | | | | |
| **Part 3** (4 items) number of errors Check if 3 or more errors. | | | | | | | | | | | |
| **Part 4** (6 items) number of errors Check if 3 or more errors. | | | | | | | | | | | |
| **Part 5** (5 items) number of errors Check if 3 or more rows. | | | | | | | | | | | |
| **Part 6** (5 items) number of errors Check if 3 or more equations. | | | | | | | | | | | |
| **Part 7** (4 items) number of errors Check if 2 or more errors. | | | | | | | | | | | |
| **Part 8** (4 items) number of errors Check if 2 or more errors. | | | | | | | | | | | |
| **Number of Parts Failed** (total checks) | | | | | | | | | | | |
| 3 or more parts failed: **Place at Lesson 1.** | | | | | | | | | | | |
| **Total Errors in Section B** | | | | | | | | | | | |
| 12 or more errors: **Place at Lesson 1.** | | | | | | | | | | | |
| Fail 2 or fewer parts with 11 or fewer errors: **Place at Lesson 16.** | | | | | | | | | | | |

# Essentials for Algebra Placement Test

## Section A

**Score**

**Name** _____

---

**Part 1** ▷ **Answer each question.**

| 5204 |

   **a.** How many digits?     _____

   **b.** What's the tens digit?     _____

   **c.** What's the thousands digit?     _____

| 217 |

   **d.** How many digits?     _____

   **e.** What's the ones digit?     _____

   **f.** What's the hundreds digit?     _____

◆ **For items g through k, circle each even number.**

   **g.** 127      **h.** 48      **i.** 150      **j.** 483      **k.** 536

◆ **Work items l through u.**

   **l.** $4 \times \boxed{\phantom{0}} = 20$      **m.** $6 \times \boxed{\phantom{0}} = 48$      **n.** $9 \times \boxed{\phantom{0}} = 54$

   **o.** $7 \times \boxed{\phantom{0}} = 21$      **p.** $6\overline{)42}$      **q.** $4\overline{)28}$

   **r.** $9\overline{)27}$      **s.** $8\overline{)64}$      **t.** $7\overline{)63}$

   **u.** $2\overline{)12}$

# Section A (continued)

**Score**

(Number missed)

**Name** _____

---

| Part 2 | Work each item. |
|---|---|

**a.** 
$$\begin{array}{r} 125 \\ \times\phantom{0}6 \\ \hline \end{array}$$

**b.** 
$$\begin{array}{r} 3300 \\ -\phantom{0}502 \\ \hline \end{array}$$

**c.** 
$$\begin{array}{r} 14 \\ 125 \\ +\phantom{0}93 \\ \hline \end{array}$$

**d.** 
$$\begin{array}{r} 28 \\ \times\phantom{0}13 \\ \hline \end{array}$$

**e.** $3\overline{)963}$

**f.** $2\overline{)758}$

**g.** $4\overline{)1208}$

**h.** 
$$\begin{array}{r} \frac{11}{4} \\ -\frac{10}{4} \\ \hline \end{array}$$

**i.** 
$$\begin{array}{r} \frac{4}{5} \\ +\frac{2}{5} \\ \hline \end{array}$$

**j.** $\dfrac{1}{8} \times \dfrac{5}{9} =$

**k.** $\dfrac{7}{5}\left(\dfrac{6}{5}\right) =$

# *Essentials for Algebra* Placement Test

## Section B

**Score**

**Name** _____

---

**Part 1** ▷ **Round each value to the nearest whole number. Then round each value to the nearest hundredth.**

39.725    **a.** _____        **b.** _____

6.2489    **c.** _____        **d.** _____

0.513    **e.** _____        **f.** _____

---

**Part 2** ▷ **Write each problem in a column and work it.**

**a.** $5.4 - 0.205$      **b.** $104.2 + 9.75$      **c.** $19 + 12.5 + 6.54$

**d.** $6 - 4.15$      **e.** $5.2 \times .003$      **f.** $142 \times .7$

# Section B (continued)

**Score**

(Number missed)

**Name** _____

---

**Part 3** Write the abbreviation for each unit name.

a. centimeters _____     c. kilometer _____

b. inches _____     d. feet _____

---

**Part 4** Find the area and perimeter of each figure.

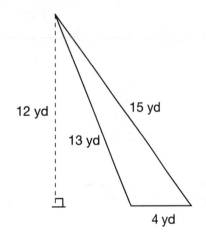

15 yd

12 yd

13 yd

4 yd

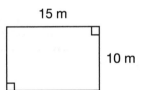

15 m

10 m

a. Area = _____

b. Perimeter = _____

e. Area = _____

f. Perimeter = _____

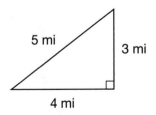

5 mi

3 mi

4 mi

c. Area = _____

d. Perimeter = _____

# Section B (continued)

**Part 5** Complete each row to show the equivalent decimal, fraction, and percent.

|   | Decimal | Fraction | Percent |
|---|---------|----------|---------|
| **a.** | 4.02 | | |
| **b.** | | $\frac{8}{100}$ | |
| **c.** | | | 70% |
| **d.** | | | 5% |
| **e.** | | $\frac{100}{100}$ | |

**Part 6** Complete each equation to show the equivalent fractions.

a. $5 = \dfrac{\square}{6} = \dfrac{\square}{1}$

b. $7 = \dfrac{\square}{2} = \dfrac{\square}{5}$

c. $\dfrac{3}{4} = \dfrac{\square}{20}$

d. $\dfrac{1}{5} = \dfrac{63}{\square}$

e. $\dfrac{2}{7} = \dfrac{14}{\square}$

## Section B (continued)

**Score**

Name _____

☐

(Number missed)

---

**Part 7** ▷ **Complete each equation to show a fraction and a mixed number.**

    **a.** $3\frac{5}{9} =$        **b.** $1\frac{7}{8} =$        **c.** $\frac{25}{3} =$        **d.** $\frac{19}{5} =$

---

**Part 8** ▷ **Work each problem.**

$15      $29.50      $12.99      $18

- **The price tags show the cost of each item.**

  **a.** Briana bought shoes and a camera.
     How much did she spend?

  **b.** Roger bought one of everything
     except the shoes. How much did
     he spend?

  **c.** Maria had $40. She bought the shoes.
     How much did she have left?

  **d.** Kahlil bought the hat and the gloves.
     He gave the clerk $30.
     How much change did he get back?

## Section A

Score

Name _____

[ ]

(Number missed)

**Part 1** Answer each question.

5204
a. How many digits? _4_
b. What's the tens digit? _0_
c. What's the thousands digit? _5_

217
d. How many digits? _3_
e. What's the ones digit? _7_
f. What's the hundreds digit? _2_

◆ For items g through k, circle each even number.

g. 127   h. (48)   i. (150)   j. 483   k. (536)

◆ Work items l through u.

l. 4 × [5] = 20   m. 6 × [8] = 48   n. 9 × [6] = 54

o. 7 × [3] = 21   p. 6)42 → 7   q. 4)28 → 7

r. 9)27 → 3   s. 8)64 → 8   t. 7)63 → 9

u. 2)12 → 6

---

Score

Name _____

[ ]

(Number missed)

**Part 2** Work each item.

a.
```
  125
×   6
 750
```

b.
```
 3300
- 502
2798
```

c.
```
  14
 125
+ 93
 232
```

d.
```
   28
×  13
   84
  280
  364
```

e. 3)963 → 321

f. 2)758 → 379

g. 4)1208 → 302

h. $\frac{11}{4} - \frac{10}{4} = \frac{1}{4}$

i. $\frac{4}{5} + \frac{2}{5} = \frac{6}{5}$ $\left[\text{or } 1\frac{1}{5}\right]$

j. $\frac{1}{8} \times \frac{5}{9} = \frac{5}{72}$

k. $\frac{7}{5}\left(\frac{6}{5}\right) = \frac{42}{25}$ $\left[\text{or } 1\frac{17}{25}\right]$

---

## Section B

Score

Name _____

[ ]

(Number missed)

**Part 1** Round each value to the nearest whole number. Then round each value to the nearest hundredth.

39.725   a. _40_   b. _39.73_

6.2489   c. _6_   d. _6.25_

0.513   e. _1_   f. _0.51_

**Part 2** Write each problem in a column and work it.

a. 5.4 − 0.205
```
 5.400
-0.205
 5.195
```

b. 104.2 + 9.75
```
104.2
+ 9.75
113.95
```

c. 19 + 12.5 + 6.54
```
  19
 12.5
+6.54
38.04
```

d. 6 − 4.15
```
 6.00
-4.15
 1.85
```

e. 5.2 × .003
```
  5.2
×.003
.0156
```

f. 142 × .7
```
 142
× .7
99.4
```

---

Score

Name _____

[ ]

(Number missed)

**Part 3** Write the abbreviation for each unit name.

a. centimeters _cm_   c. kilometer _km_

b. inches _in._   d. feet _ft_

**Part 4** Find the area and perimeter of each figure.

a. Area = _150 sq m_
b. Perimeter = _50 m_

e. Area = _24 sq yd_
f. Perimeter = _32 yd_

c. Area = _6 sq mi_
d. Perimeter = _12 mi_

---

**Score**

[ ]

(Number missed)

Name _____

**Part 5** Complete each row to show the equivalent decimal, fraction, and percent.

| | Decimal | Fraction | Percent |
|---|---|---|---|
| a. | 4.02 | $\frac{402}{100}$ | 402% |
| b. | .08 | $\frac{8}{100}$ | 8% |
| c. | .70 | $\frac{70}{100}$ | 70% |
| d. | .05 | $\frac{5}{100}$ | 5% |
| e. | 1.00 | $\frac{100}{100}$ | 100% |

**Part 6** Complete each equation to show the equivalent fractions.

a. $5 = \dfrac{\boxed{30}}{6} = \dfrac{\boxed{5}}{1}$

b. $7 = \dfrac{\boxed{14}}{2} = \dfrac{\boxed{35}}{5}$

c. $\dfrac{3}{4} = \dfrac{\boxed{15}}{20}$

d. $\dfrac{1}{5} = \dfrac{63}{\boxed{315}}$

e. $\dfrac{2}{7} = \dfrac{14}{\boxed{49}}$

**Score**

[ ]

(Number missed)

Name _____

**Part 7** Complete each equation to show a fraction and a mixed number.

a. $3\frac{5}{9} = \dfrac{32}{9}$  b. $1\frac{7}{8} = \dfrac{15}{8}$  c. $\dfrac{25}{3} = 8\frac{1}{3}$  d. $\dfrac{19}{5} = 3\frac{4}{5}$

**Part 8** Work each problem.

$15  $29.50  $12.99  $18

- The price tags show the cost of each item.

a. Briana bought shoes and a camera. How much did she spend?

$29.50
+ 18
$47.50

b. Roger bought one of everything except the shoes. How much did he spend?

$15
12.99
+ 18
$45.99

c. Maria had $40. She bought the shoes. How much did she have left?

$40.00
− 29.50
$10.50

d. Kahlil bought the hat and the gloves. He gave the clerk $30. How much change did he get back?

$15         $30.00
+ 12.99     − 27.99
$27.99      $ 2.01

# Remedy Summary—Group Summary of Mastery Test Performance

**Note:** Test remedies are specified in the Answer Key. Percent Summary is also specified in the Answer Key.

| Name | Test 1A (Use after lesson 6.) Check parts not passed. | | | | | | | Total % | Test 1B (Use after lesson 15.) Check parts not passed. | | | | | | | | | | Total % |
|---|---|---|---|---|---|---|---|---|---|---|---|---|---|---|---|---|---|---|---|
| | 1 | 2 | 3 | 4 | 5 | 6 | 7 | | 1 | 2 | 3 | 4 | 5 | 6 | 7 | 8 | 9 | 10 | |
| 1. | | | | | | | | | | | | | | | | | | | |
| 2. | | | | | | | | | | | | | | | | | | | |
| 3. | | | | | | | | | | | | | | | | | | | |
| 4. | | | | | | | | | | | | | | | | | | | |
| 5. | | | | | | | | | | | | | | | | | | | |
| 6. | | | | | | | | | | | | | | | | | | | |
| 7. | | | | | | | | | | | | | | | | | | | |
| 8. | | | | | | | | | | | | | | | | | | | |
| 9. | | | | | | | | | | | | | | | | | | | |
| 10. | | | | | | | | | | | | | | | | | | | |
| 11. | | | | | | | | | | | | | | | | | | | |
| 12. | | | | | | | | | | | | | | | | | | | |
| 13. | | | | | | | | | | | | | | | | | | | |
| 14. | | | | | | | | | | | | | | | | | | | |
| 15. | | | | | | | | | | | | | | | | | | | |
| 16. | | | | | | | | | | | | | | | | | | | |
| 17. | | | | | | | | | | | | | | | | | | | |
| 18. | | | | | | | | | | | | | | | | | | | |
| 19. | | | | | | | | | | | | | | | | | | | |
| 20. | | | | | | | | | | | | | | | | | | | |
| 21. | | | | | | | | | | | | | | | | | | | |
| 22. | | | | | | | | | | | | | | | | | | | |
| 23. | | | | | | | | | | | | | | | | | | | |
| 24. | | | | | | | | | | | | | | | | | | | |
| 25. | | | | | | | | | | | | | | | | | | | |
| 26. | | | | | | | | | | | | | | | | | | | |
| 27. | | | | | | | | | | | | | | | | | | | |
| 28. | | | | | | | | | | | | | | | | | | | |
| 29. | | | | | | | | | | | | | | | | | | | |
| 30. | | | | | | | | | | | | | | | | | | | |
| Number of students not passed = NP | | | | | | | | | | | | | | | | | | | |
| Total number of students = T | | | | | | | | | | | | | | | | | | | |
| Remedy needed if NP/T = 25% or more | | | | | | | | | | | | | | | | | | | |

# Remedy Summary—Group Summary of Mastery Test Performance

*Note:* Test remedies are specified in the *Answer Key.*
Percent Summary is also specified in the *Answer Key.*

| Name | Test 2 — Check parts not passed. | | | | | | | | | Total % | Test 3 — Check parts not passed. | | | | | | | | | Total % |
|---|---|---|---|---|---|---|---|---|---|---|---|---|---|---|---|---|---|---|---|---|
| | 1 | 2 | 3 | 4 | 5 | 6 | 7 | 8 | 9 | | 1 | 2 | 3 | 4 | 5 | 6 | 7 | 8 | 9 | |
| 1. | | | | | | | | | | | | | | | | | | | | |
| 2. | | | | | | | | | | | | | | | | | | | | |
| 3. | | | | | | | | | | | | | | | | | | | | |
| 4. | | | | | | | | | | | | | | | | | | | | |
| 5. | | | | | | | | | | | | | | | | | | | | |
| 6. | | | | | | | | | | | | | | | | | | | | |
| 7. | | | | | | | | | | | | | | | | | | | | |
| 8. | | | | | | | | | | | | | | | | | | | | |
| 9. | | | | | | | | | | | | | | | | | | | | |
| 10. | | | | | | | | | | | | | | | | | | | | |
| 11. | | | | | | | | | | | | | | | | | | | | |
| 12. | | | | | | | | | | | | | | | | | | | | |
| 13. | | | | | | | | | | | | | | | | | | | | |
| 14. | | | | | | | | | | | | | | | | | | | | |
| 15. | | | | | | | | | | | | | | | | | | | | |
| 16. | | | | | | | | | | | | | | | | | | | | |
| 17. | | | | | | | | | | | | | | | | | | | | |
| 18. | | | | | | | | | | | | | | | | | | | | |
| 19. | | | | | | | | | | | | | | | | | | | | |
| 20. | | | | | | | | | | | | | | | | | | | | |
| 21. | | | | | | | | | | | | | | | | | | | | |
| 22. | | | | | | | | | | | | | | | | | | | | |
| 23. | | | | | | | | | | | | | | | | | | | | |
| 24. | | | | | | | | | | | | | | | | | | | | |
| 25. | | | | | | | | | | | | | | | | | | | | |
| 26. | | | | | | | | | | | | | | | | | | | | |
| 27. | | | | | | | | | | | | | | | | | | | | |
| 28. | | | | | | | | | | | | | | | | | | | | |
| 29. | | | | | | | | | | | | | | | | | | | | |
| 30. | | | | | | | | | | | | | | | | | | | | |
| Number of students not passed = NP | | | | | | | | | | | | | | | | | | | | |
| Total number of students = T | | | | | | | | | | | | | | | | | | | | |
| Remedy needed if NP/T = 25% or more | | | | | | | | | | | | | | | | | | | | |

# Remedy Summary—Group Summary of Mastery Test Performance

**Note:** Test remedies are specified in the *Answer Key*. Percent Summary is also specified in the *Answer Key*.

| Name | Test 4 | | | | | | | | | | | Test 5 | | | | | | | | | | | | | |
|---|---|---|---|---|---|---|---|---|---|---|---|---|---|---|---|---|---|---|---|---|---|---|---|---|---|
| | 1 | 2 | 3 | 4 | 5 | 6 | 7 | 8 | 9 | 10 | Total % | 1 | 2 | 3 | 4 | 5 | 6 | 7 | 8 | 9 | 10 | 11 | 12 | 13 | Total % |
| | Check parts not passed. | | | | | | | | | | | Check parts not passed. | | | | | | | | | | | | |
| 1. | | | | | | | | | | | | | | | | | | | | | | | | | |
| 2. | | | | | | | | | | | | | | | | | | | | | | | | | |
| 3. | | | | | | | | | | | | | | | | | | | | | | | | | |
| 4. | | | | | | | | | | | | | | | | | | | | | | | | | |
| 5. | | | | | | | | | | | | | | | | | | | | | | | | | |
| 6. | | | | | | | | | | | | | | | | | | | | | | | | | |
| 7. | | | | | | | | | | | | | | | | | | | | | | | | | |
| 8. | | | | | | | | | | | | | | | | | | | | | | | | | |
| 9. | | | | | | | | | | | | | | | | | | | | | | | | | |
| 10. | | | | | | | | | | | | | | | | | | | | | | | | | |
| 11. | | | | | | | | | | | | | | | | | | | | | | | | | |
| 12. | | | | | | | | | | | | | | | | | | | | | | | | | |
| 13. | | | | | | | | | | | | | | | | | | | | | | | | | |
| 14. | | | | | | | | | | | | | | | | | | | | | | | | | |
| 15. | | | | | | | | | | | | | | | | | | | | | | | | | |
| 16. | | | | | | | | | | | | | | | | | | | | | | | | | |
| 17. | | | | | | | | | | | | | | | | | | | | | | | | | |
| 18. | | | | | | | | | | | | | | | | | | | | | | | | | |
| 19. | | | | | | | | | | | | | | | | | | | | | | | | | |
| 20. | | | | | | | | | | | | | | | | | | | | | | | | | |
| 21. | | | | | | | | | | | | | | | | | | | | | | | | | |
| 22. | | | | | | | | | | | | | | | | | | | | | | | | | |
| 23. | | | | | | | | | | | | | | | | | | | | | | | | | |
| 24. | | | | | | | | | | | | | | | | | | | | | | | | | |
| 25. | | | | | | | | | | | | | | | | | | | | | | | | | |
| 26. | | | | | | | | | | | | | | | | | | | | | | | | | |
| 27. | | | | | | | | | | | | | | | | | | | | | | | | | |
| 28. | | | | | | | | | | | | | | | | | | | | | | | | | |
| 29. | | | | | | | | | | | | | | | | | | | | | | | | | |
| 30. | | | | | | | | | | | | | | | | | | | | | | | | | |
| Number of students not passed = NP | | | | | | | | | | | | | | | | | | | | | | | | | |
| Total number of students = T | | | | | | | | | | | | | | | | | | | | | | | | | |
| Remedy needed if NP/T = 25% or more | | | | | | | | | | | | | | | | | | | | | | | | | |

# Remedy Summary—Group Summary of Mastery Test Performance

**Note:** Test remedies are specified in the *Answer Key*. Percent Summary is also specified in the *Answer Key*.

## Test 6

Check parts not passed.

| Name | 1 | 2 | 3 | 4 | 5 | 6 | 7 | 8 | 9 | 10 | 11 | 12 | Total % |
|---|---|---|---|---|---|---|---|---|---|---|---|---|---|
| 1. | | | | | | | | | | | | | |
| 2. | | | | | | | | | | | | | |
| 3. | | | | | | | | | | | | | |
| 4. | | | | | | | | | | | | | |
| 5. | | | | | | | | | | | | | |
| 6. | | | | | | | | | | | | | |
| 7. | | | | | | | | | | | | | |
| 8. | | | | | | | | | | | | | |
| 9. | | | | | | | | | | | | | |
| 10. | | | | | | | | | | | | | |
| 11. | | | | | | | | | | | | | |
| 12. | | | | | | | | | | | | | |
| 13. | | | | | | | | | | | | | |
| 14. | | | | | | | | | | | | | |
| 15. | | | | | | | | | | | | | |
| 16. | | | | | | | | | | | | | |
| 17. | | | | | | | | | | | | | |
| 18. | | | | | | | | | | | | | |
| 19. | | | | | | | | | | | | | |
| 20. | | | | | | | | | | | | | |
| 21. | | | | | | | | | | | | | |
| 22. | | | | | | | | | | | | | |
| 23. | | | | | | | | | | | | | |
| 24. | | | | | | | | | | | | | |
| 25. | | | | | | | | | | | | | |
| 26. | | | | | | | | | | | | | |
| 27. | | | | | | | | | | | | | |
| 28. | | | | | | | | | | | | | |
| 29. | | | | | | | | | | | | | |
| 30. | | | | | | | | | | | | | |

## Test 7

Check parts not passed.

| 1 | 2 | 3 | 4 | 5 | 6 | 7 | 8 | 9 | 10 | 11 | 12 | 13 | 14 | 15 | Total % |
|---|---|---|---|---|---|---|---|---|---|---|---|---|---|---|---|

Number of students not passed = NP

Total number of students = T

Remedy needed if NP/T = 25% or more

# Remedy Summary—Group Summary of Mastery Test Performance

Copyright SRA/McGraw-Hill. Permission is granted to reproduce this page for classroom use.

**Note:** Test remedies are specified in the *Answer Key*. Percent Summary is also specified in the *Answer Key*.

## Test 8

Check parts not passed.

| Name | 1 | 2 | 3 | 4 | 5 | 6 | 7 | 8 | 9 | 10 | 11 | 12 | 13 | Total % |
|------|---|---|---|---|---|---|---|---|---|----|----|----|----|---------|
| 1. | | | | | | | | | | | | | | |
| 2. | | | | | | | | | | | | | | |
| 3. | | | | | | | | | | | | | | |
| 4. | | | | | | | | | | | | | | |
| 5. | | | | | | | | | | | | | | |
| 6. | | | | | | | | | | | | | | |
| 7. | | | | | | | | | | | | | | |
| 8. | | | | | | | | | | | | | | |
| 9. | | | | | | | | | | | | | | |
| 10. | | | | | | | | | | | | | | |
| 11. | | | | | | | | | | | | | | |
| 12. | | | | | | | | | | | | | | |
| 13. | | | | | | | | | | | | | | |
| 14. | | | | | | | | | | | | | | |
| 15. | | | | | | | | | | | | | | |
| 16. | | | | | | | | | | | | | | |
| 17. | | | | | | | | | | | | | | |
| 18. | | | | | | | | | | | | | | |
| 19. | | | | | | | | | | | | | | |
| 20. | | | | | | | | | | | | | | |
| 21. | | | | | | | | | | | | | | |
| 22. | | | | | | | | | | | | | | |
| 23. | | | | | | | | | | | | | | |
| 24. | | | | | | | | | | | | | | |
| 25. | | | | | | | | | | | | | | |
| 26. | | | | | | | | | | | | | | |
| 27. | | | | | | | | | | | | | | |
| 28. | | | | | | | | | | | | | | |
| 29. | | | | | | | | | | | | | | |
| 30. | | | | | | | | | | | | | | |

Number of students not passed = NP

Total number of students = T

Remedy needed if NP/T = 25% or more

## Test 9

Check parts not passed.

| 1 | 2 | 3 | 4 | 5 | 6 | 7 | 8 | 9 | 10 | 11 | 12 | 13 | 14 | 15 | Total % |
|---|---|---|---|---|---|---|---|---|----|----|----|----|----|----|---------|

**172** *Remedy Summary*

# Remedy Summary—Group Summary of Mastery Test Performance

**Note:** Test remedies are specified in the *Answer Key*. Percent Summary is also specified in the *Answer Key*.

## Test 10

Check parts not passed.

| Name | 1 | 2 | 3 | 4 | 5 | 6 | 7 | 8 | 9 | 10 | Total % |
|---|---|---|---|---|---|---|---|---|---|---|---|
| 1. | | | | | | | | | | | |
| 2. | | | | | | | | | | | |
| 3. | | | | | | | | | | | |
| 4. | | | | | | | | | | | |
| 5. | | | | | | | | | | | |
| 6. | | | | | | | | | | | |
| 7. | | | | | | | | | | | |
| 8. | | | | | | | | | | | |
| 9. | | | | | | | | | | | |
| 10. | | | | | | | | | | | |
| 11. | | | | | | | | | | | |
| 12. | | | | | | | | | | | |
| 13. | | | | | | | | | | | |
| 14. | | | | | | | | | | | |
| 15. | | | | | | | | | | | |
| 16. | | | | | | | | | | | |
| 17. | | | | | | | | | | | |
| 18. | | | | | | | | | | | |
| 19. | | | | | | | | | | | |
| 20. | | | | | | | | | | | |
| 21. | | | | | | | | | | | |
| 22. | | | | | | | | | | | |
| 23. | | | | | | | | | | | |
| 24. | | | | | | | | | | | |
| 25. | | | | | | | | | | | |
| 26. | | | | | | | | | | | |
| 27. | | | | | | | | | | | |
| 28. | | | | | | | | | | | |
| 29. | | | | | | | | | | | |
| 30. | | | | | | | | | | | |
| Number of students not passed = NP | | | | | | | | | | | |
| Total number of students = T | | | | | | | | | | | |
| Remedy needed if NP/T = 25% or more | | | | | | | | | | | |

## Test 11

Check parts not passed.

| Name | 1 | 2 | 3 | 4 | 5 | 6 | 7 | 8 | 9 | 10 | 11 | 12 | 13 | 14 | 15 | Total % |
|---|---|---|---|---|---|---|---|---|---|---|---|---|---|---|---|---|
| 1. | | | | | | | | | | | | | | | | |
| 2. | | | | | | | | | | | | | | | | |
| 3. | | | | | | | | | | | | | | | | |
| 4. | | | | | | | | | | | | | | | | |
| 5. | | | | | | | | | | | | | | | | |
| 6. | | | | | | | | | | | | | | | | |
| 7. | | | | | | | | | | | | | | | | |
| 8. | | | | | | | | | | | | | | | | |
| 9. | | | | | | | | | | | | | | | | |
| 10. | | | | | | | | | | | | | | | | |
| 11. | | | | | | | | | | | | | | | | |
| 12. | | | | | | | | | | | | | | | | |
| 13. | | | | | | | | | | | | | | | | |
| 14. | | | | | | | | | | | | | | | | |
| 15. | | | | | | | | | | | | | | | | |
| 16. | | | | | | | | | | | | | | | | |
| 17. | | | | | | | | | | | | | | | | |
| 18. | | | | | | | | | | | | | | | | |
| 19. | | | | | | | | | | | | | | | | |
| 20. | | | | | | | | | | | | | | | | |
| 21. | | | | | | | | | | | | | | | | |
| 22. | | | | | | | | | | | | | | | | |
| 23. | | | | | | | | | | | | | | | | |
| 24. | | | | | | | | | | | | | | | | |
| 25. | | | | | | | | | | | | | | | | |
| 26. | | | | | | | | | | | | | | | | |
| 27. | | | | | | | | | | | | | | | | |
| 28. | | | | | | | | | | | | | | | | |
| 29. | | | | | | | | | | | | | | | | |
| 30. | | | | | | | | | | | | | | | | |

## Part 1

Write the abbreviation
for each unit name.

| Unit Name | Abbreviation |
|---|---|
| miles | _____ |
| centimeters | _____ |
| yards | _____ |
| meters | _____ |
| feet | _____ |
| kilometers | _____ |
| inches | _____ |

## Part 2

Cross out the problems you **cannot** work the way
they are written. Work the problems you **can** work
the way they are written.

**a.** $4 - \frac{3}{5} =$ ☐

**b.** $\frac{9}{12} - \frac{7}{12} =$ ☐

**c.** $\frac{20}{m} + \frac{7}{m} =$ ☐

**d.** $\frac{11}{c} + \frac{11}{d} =$ ☐

**e.** $1 - \frac{5}{9} =$ ☐

**f.** $\frac{13}{35} + \frac{4}{35} =$ ☐

**g.** $\frac{3}{5} + \frac{5}{3} =$ ☐

**h.** $\frac{10}{2n} - \frac{9}{2n} =$ ☐

## Lesson 3

## Part 1

Cross out the problems you **cannot** work the way they are written.
Work the problems you **can** work the way they are written.

**a.** $\frac{2}{3} \times \frac{1}{5} =$ ☐

**b.** $\frac{2}{3} + \frac{1}{5} =$ ☐

**c.** $\frac{8}{p} + \frac{17}{d} =$ ☐

**d.** $\frac{12}{9} \times \frac{2}{7} =$ ☐

**e.** $\frac{10}{1} \times \frac{3}{8} =$ ☐

**f.** $10 - \frac{3}{8} =$ ☐

**g.** $\frac{1}{20} \times \frac{23}{3} =$ ☐

**Part 2** Decimal Values for Hundredths

$$\frac{4}{100} = .04 \qquad \frac{536}{100} = 5.36$$

◆ **Complete the table.**

|   | Decimal | Fraction | Percent |
|---|---------|----------|---------|
|   | 2.06 | $\frac{206}{100}$ | 206% |
| a. |   | $\frac{186}{100}$ |   |
| b. |   | $\frac{80}{100}$ |   |
| c. |   | $\frac{7}{100}$ |   |
| d. |   | $\frac{15}{100}$ |   |
| e. |   | $\frac{258}{100}$ |   |

**Part 2** Complete the table.

|   | Decimal | Fraction | Percent |
|---|---------|----------|---------|
| a. |   | $\frac{100}{100}$ |   |
| b. |   |   | 802% |
| c. | 4.03 |   |   |
| d. |   |   | 8% |
| e. |   | $\frac{90}{100}$ |   |

Part 1 > Write an addition equation for each mixed number.

a. $3\frac{9}{10}$

d. $4\frac{5}{7}$

_____

_____

b. $25\frac{1}{2}$

e. $20\frac{5}{6}$

_____

_____

c. $8\frac{1}{3}$

f. $7\frac{3}{8}$

_____

_____

**Part 1** > Changing Both Sides of an Equation

◆ If you change **one side**, you must change the
  other side in the **same way.**

◆ For each item, change both sides of the equation in the same way.
  Figure out what the letter equals.

a. $k - 2 = 23$

d. $t - 30 = 1$

b. $17 + r = 24$

e. $\frac{3}{5} + y = \frac{5}{5}$

c. $g + 13 = 17$

f. $m - 100 = 256$

**Lesson 30**

Part 3 > Make a point on the coordinate system for each description. Label each point.

**Point A**  $x = 5$,  $y = 7$
**Point B**  $x = 8$,  $y = 10$
**Point C**  $x = 0$,  $y = 2$
**Point D**  $x = 3$,  $y = 5$

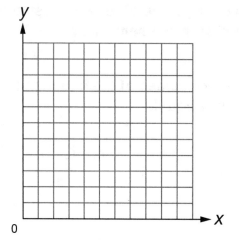

**Lesson 31**

Part 1 > Make a point on the coordinate system for each description. Label each point.

**Point A**  $x = 10$,  $y = 6$
**Point B**  $x = 5$,  $y = 1$
**Point C**  $x = 8$,  $y = 4$
**Point D**  $x = 9$,  $y = 5$

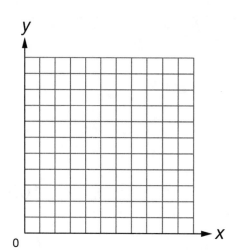

**178**  *Remedies*

**Lesson 37**

Part 1 > Complete the table. Then plot 3 points and draw the line.
Then plot the remaining points.

| Function | | | |
|---|---|---|---|
| $x$ | $+3$ | $=$ | $y$ |
| **A** | 6 | | |
| **B** | 9 | | |
| **C** | 0 | | |
| **D** | 2 | | |
| **E** | 7 | | |
| **F** | 4 | | |

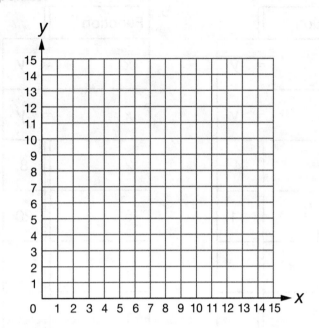

**Lesson 41**

Part 2 > Label the axes. Write the description for points A–E.
Plot points F–J on the coordinate system.

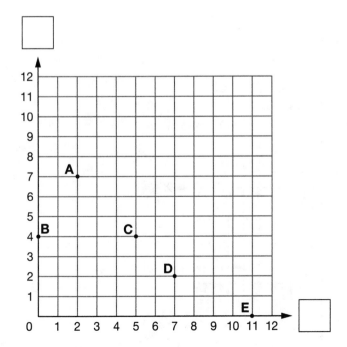

**F** (7, 11)

**G** (2, 5)

**H** (0, 9)

**I** (12, 1)

**J** (3, 0)

**Part 2** ▷ Complete each function table.

**a.**

| Function | |
|---|---|
| $x$      = $y$ | |
| $x$      = $y$ | |
| 7 | 4 |
| 14 | 11 |
| 3 | |
| 6 | |
| 10 | |

**b.**

| Function | |
|---|---|
| $x + 6$ = $y$ | |
| $x\ (4)$ = $y$ | ✓ |
| 2 | 8 |
| 5 | 20 |
| 3 | |
| 1 | |
| 6 | |

**c.**

| Function | |
|---|---|
| $x + 2$ = $y$ | |
| $x\ (1,25)$ = $y$ | |
| 8 | 10 |
| 4 | 5 |
| 12 | 15 |
| 16 | 20 |
| 0 | 0 |

**Lesson 44**

**Part 3** ▷ Write the coordinates next to each point. You do not need a plus sign for positive $x$ or $y$ values.

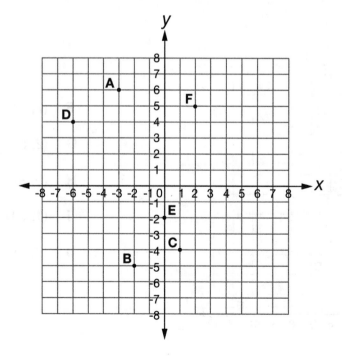

$4X = \dfrac{6}{4}$

$8X = \dfrac{10}{8}$

## Lesson 47

**Part 1** > Combining Letter Terms with Fractions

$$7v + \frac{2}{3}v = 5 \qquad\qquad v + \frac{2}{3}v = 5$$

$$\frac{21}{3}v + \frac{2}{3}v = 5 \qquad\qquad \frac{3}{3}v + \frac{2}{3}v = 5$$

$$\frac{23}{3}v = 5 \qquad\qquad \frac{5}{3}v = 5$$

◆ **Combine like terms. Solve for the letter.**

**a.** $\frac{9}{5}v + v = 28$    **b.** $3j - \frac{1}{2}j + 11 - 10 = 21$    **c.** $0 = r + 2 - 6 - \frac{1}{5}r$

$$4x$$
$$\frac{4x = 3}{4} \quad \frac{}{3\!\!\!/4}$$

---

**Part 2** > Complete each table.

**1.**

| Function | | |
|---|---|---|
| $x$ $\left(\frac{y}{x}\right) = y$ | | |
| $x$ $\left(\frac{3}{4}\right) = y$ | | |
| **A** | 4 | 3 |
| **B** | 28 | 21 |
| **C** | 20 | |

**2.**

| Function | | |
|---|---|---|
| $x$ $\left(\frac{y}{x}\right) = y$ | | |
| $x$ $(\ ) = y$ | | |
| **A** | 2 | 6 |
| **B** | 12 | |
| **C** | | 9 |
| **D** | 1 | |

**3.**

| Function | | |
|---|---|---|
| $x$ $\left(\frac{y}{x}\right) = y$ | | |
| $x$ $(\ ) = y$ | | |
| **A** | 9 | 3 |
| **B** | | 5 |
| **C** | 6 | |

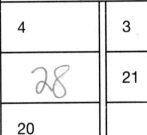

**Part 1** Complete the table. Plot points A through D. Draw the line.

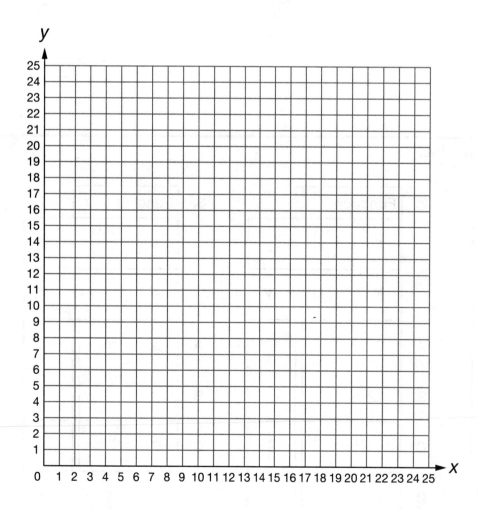

| | $y = \frac{3}{5} x$ | |
|---|---|---|
| **A** | 6 | |
| **B** | | 20 |
| **C** | | 15 |
| **D** | 15 | |

$y = \frac{3}{5}x + 1$

$y + x = \frac{3}{5}$

$y + x = \frac{3}{5}$

$6 + x = \frac{3}{5}$

$\frac{3}{5} - \frac{x}{1}$

$\frac{3}{5} - \frac{20}{5}$

$\frac{17}{5}$

Part 2 $\rangle$ **Rewrite each fraction. First show a base with a positive exponent.**
**Then show a base with a negative exponent.**

**Sample** $\dfrac{4 \times 4 \times 4}{4 \times 4 \times 4 \times 4 \times 4} = \dfrac{1}{4^2} = 4^{-2}$

**a.** $\dfrac{7 \times 7 \times 7 \times 7 \times 7 \times 7}{7 \times 7} = \boxed{\phantom{xx}} = \boxed{\phantom{xx}}$

**b.** $\dfrac{8 \times 8 \times 8}{8 \times 8 \times 8 \times 8 \times 8 \times 8} = \boxed{\phantom{xx}} = \boxed{\phantom{xx}}$

**c.** $\dfrac{10}{10 \times 10 \times 10 \times 10 \times 10 \times 10} = \boxed{\phantom{xx}} = \boxed{\phantom{xx}}$

**d.** $\dfrac{m \times m \times m \times m \times m \times m \times m \times m}{m \times m} = \boxed{\phantom{xx}} = \boxed{\phantom{xx}}$

**Lesson 58**

Part 2 $\rangle$ $\rangle$ Order of Operations

$-2 (+4) - 1 + 8 (-2) = \blacksquare$         ◆ First remove parentheses.

$-8 \quad -1 \quad -16 \quad = \blacksquare$         ◆ Then combine.

**a.** $+6 - 2(-3) + 4(-1) - 5 = \blacksquare$          **c.** $-7(-1) + 12 - 2(-9) = \blacksquare$

$\underline{\phantom{xxxxxxxxxxxxxxxxxx}} = \boxed{\phantom{xxxx}}$          $\underline{\phantom{xxxxxxxxxxxxxxxxxx}} = \boxed{\phantom{xxxx}}$

**b.** $+2(-5) - 8 - 3(3) + 6 = \blacksquare$          **d.** $-4(+6) - 8(7) - 10 = \blacksquare$

$\underline{\phantom{xxxxxxxxxxxxxxxxxx}} = \boxed{\phantom{xxxx}}$          $\underline{\phantom{xxxxxxxxxxxxxxxxxx}} = \boxed{\phantom{xxxx}}$

**Part 1** ⟩ ⟩ | Simplifying Expressions with Exponents |

$m^8\ 5^2\ 5\ =\ \mathbf{5^3}\ m^8\ =\ \boxed{125\ m^8}$

◀ Write the number value first, then the letter value.

$k^{-1}\ 2^3\ b^{-2}\ k^2\ 2^{-7}\ =\ \mathbf{2^{-4}}\ kb^{-2}\ =\ \boxed{\dfrac{1}{16}\ kb^{-2}}$

◀ If the number has a negative exponent, write it as a fraction.

**Part 2** ⟩ Simplify. Rewrite the expression with the number base first. Then write the expression with a number that has no exponent.

**a.** $j^{-3}\ r^3\ 15^3\ r$

**b.** $7^3\ k^3\ 7^2\ k^{-2}$

**c.** $b^{-3}\ b\ 4^2\ m^{-1}\ 4^{-5}$

**d.** $g^{-1}\ n^2\ 10^3\ n\ 10^{-1}$

**e.** $3^2\ p^3\ p^{-1}\ 3^{-5}\ d^{-10}$

Part 1 〉 **Whole Number Square Roots, 1–20**

$$\begin{array}{cccccccccc} 1 & 2 & 3 & 4 & 5 & 6 & 7 & 8 & 9 & 10 \\ \sqrt{1} & \sqrt{4} & \sqrt{9} & \sqrt{16} & \sqrt{25} & \sqrt{36} & \sqrt{49} & \sqrt{64} & \sqrt{81} & \sqrt{100} \end{array}$$

$$\begin{array}{cccccccccc} 11 & 12 & 13 & 14 & 15 & 16 & 17 & 18 & 19 & 20 \\ \sqrt{121} & \sqrt{144} & \sqrt{169} & \sqrt{196} & \sqrt{225} & \sqrt{256} & \sqrt{289} & \sqrt{324} & \sqrt{361} & \sqrt{400} \end{array}$$

◆ **For each square root, circle *whole number* or *between whole numbers*. Below, write the whole number, or write the 2 numbers the value is between.**

a. $\sqrt{64}$      whole number      between whole numbers

_____      _____

b. $\sqrt{111}$      whole number      between whole numbers

_____      _____

c. $\sqrt{275}$      whole number      between whole numbers

_____      _____

d. $\sqrt{83}$      whole number      between whole numbers

_____      _____

e. $\sqrt{144}$      whole number      between whole numbers

_____      _____

**Lesson 80**

**Test 8–BLM 2**

**Part 1** ◇ **Label the x and y axes. Then make a line for each equation.**

**A** $y = 3x - 4$

**B** $y = -x + 5$

**C** $y = -\dfrac{3}{5}x + 2$

**D** $y = x - 8$

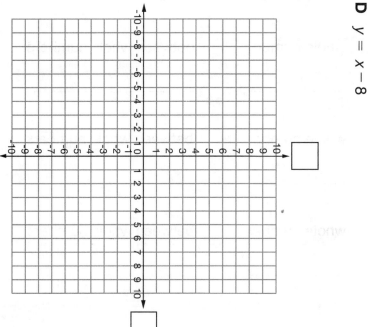

**Part 2** ⟩ **Combining Exponents**

◆ If you have a fraction with terms multiplied, move all the terms to the **numerator** before you combine exponents.

$$\frac{r^{-2}\,5^2}{r^2\,5^{-1}} = r^{-2}\,5^2\,r^{-2}\,5^1$$

$$= 5^3\,r^{-4}$$

$$= \boxed{125\,r^{-4}}$$

**Part 3** ⟩ **Simplify each expression.**

**a.** $\dfrac{f^3}{2^{-5}\,f^{-2}\,2^3} =$

$=$

$=$

**b.** $\dfrac{v\,10^{-2}}{v^{-2}\,10^2} =$

$=$

$=$

**c.** $\dfrac{m^4\,3^7}{m^{-2}\,3^9\,m^3} =$

$=$

$=$

**Part 2** ⟩   **Plot the line for each equation.**

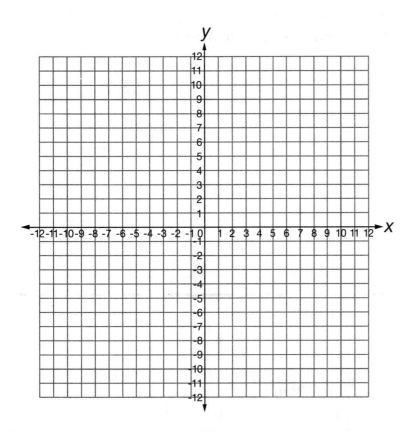

**A**   $y = \frac{1}{4}x - 2$

**B**   $y = -2x$

**C**   $y = -\frac{4}{5}x$

**D**   $y = \frac{8}{3}x + 4$

**Lesson 105**

**Part 1** Write the simplified expression below each item.

**a.** $9m^{-2} - 4m^2 + 3m^2 - 5m^{-2}$

**c.** $j^{-4} + 4j^5 - 8j^5 + 5j^{-4} - 4^2$

**b.** $r^3 - r^2 + 2m^2 + 3r^3 + 2^2$

**d.** $-3b + b^3 - 4b^3 + b^2 + 3^3 - 3^2$

**Lesson 99**

**Part 2** Scientific Notation

◆ Some numbers have only 1 digit before zeros.

$$50{,}000 = 5.0 \times 10^4$$
$$50{,}001 = 5.001 \times 10^4$$

◆ **Complete each equation.**

**a.** _____ $= 5.23 \times 10^7$

**d.** _____ $= 8.275 \times 10^2$

**b.** $1{,}040{,}000 =$ _____

**e.** _____ $= 2.0 \times 10^4$

**c.** $300{,}000 =$ _____

**f.** $800.1 =$ _____

Part 2 > For each equation, plot 2 points and draw the line.
Check where the line crosses the *y* axis.

**Line A:** $y = -\frac{1}{2}x + 3$

For one point, $x = -2$.

For another point, $y = 6$.

**Line B:** $y = 3x - 5$

For one point, $x = 1$.

For another point, $y = 4$.

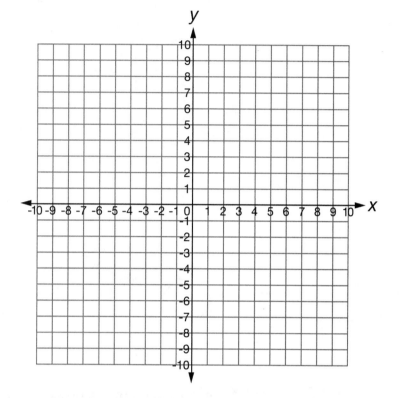

**Part 1** > Complete each equation to show a number and the equivalent scientific notation.

a. [ ] $= 3.1 \times 10^{-6}$   d. [ ] $= 8.06 \times 10^{-4}$

b. [ ] $= 0.00407$   e. [ ] $= 1.50 \times 10^{5}$

c. $90{,}000{,}000{,}000 =$ [ ]   f. $.03021 =$ [ ]

**Part 2** > Solve each pair of equations for $x$ and $y$. Plot the point for the solution. Then draw the line for each equation.

a. $y = -x + 8$

$y = 2x - 7$

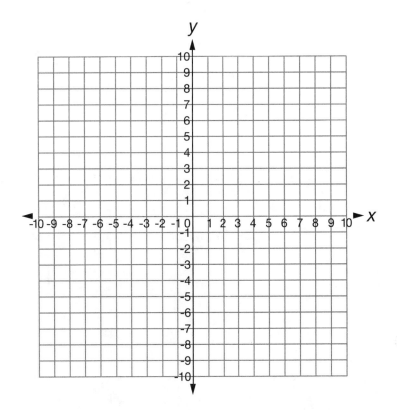

**(Part 2 continues on BLM 3.)**

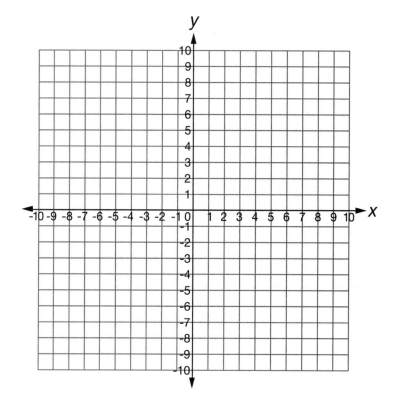

> **Part 2** Solve each pair of equations for *x* and *y*. Plot the point
> for the solution. Then draw the line for each equation.

**b.** $y = 5x + 3$

$y = 2x$

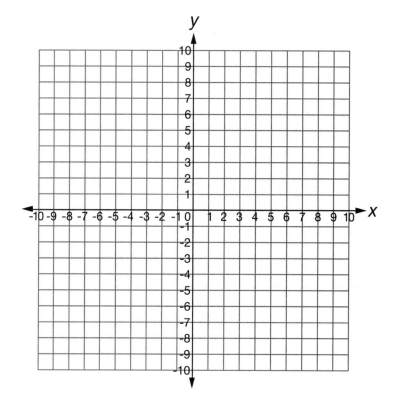

**c.** $y = x - 4$

$y = 3x - 4$

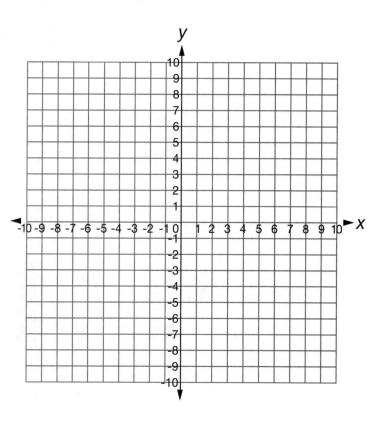

**Practice Test 1**

Name _____

**Mark the letter of the correct choice for the items shown in Practice Test 1.**

| | | | | | | | | | | |
|---|---|---|---|---|---|---|---|---|---|---|
| 1. A○ | B○ | C○ | D○ | E○ | | 31. A○ | B○ | C○ | D○ | E○ |
| 2. A○ | B○ | C○ | D○ | E○ | | 32. A○ | B○ | C○ | D○ | E○ |
| 3. A○ | B○ | C○ | D○ | E○ | | 33. A○ | B○ | C○ | D○ | E○ |
| 4. A○ | B○ | C○ | D○ | E○ | | 34. A○ | B○ | C○ | D○ | E○ |
| 5. A○ | B○ | C○ | D○ | E○ | | 35. A○ | B○ | C○ | D○ | E○ |
| 6. A○ | B○ | C○ | D○ | E○ | | 36. A○ | B○ | C○ | D○ | E○ |
| 7. A○ | B○ | C○ | D○ | E○ | | 37. A○ | B○ | C○ | D○ | E○ |
| 8. A○ | B○ | C○ | D○ | E○ | | 38. A○ | B○ | C○ | D○ | E○ |
| 9. A○ | B○ | C○ | D○ | E○ | | 39. A○ | B○ | C○ | D○ | E○ |
| 10. A○ | B○ | C○ | D○ | E○ | | 40. A○ | B○ | C○ | D○ | E○ |
| 11. A○ | B○ | C○ | D○ | E○ | | 41. A○ | B○ | C○ | D○ | E○ |
| 12. A○ | B○ | C○ | D○ | E○ | | 42. A○ | B○ | C○ | D○ | E○ |
| 13. A○ | B○ | C○ | D○ | E○ | | 43. A○ | B○ | C○ | D○ | E○ |
| 14. A○ | B○ | C○ | D○ | E○ | | 44. A○ | B○ | C○ | D○ | E○ |
| 15. A○ | B○ | C○ | D○ | E○ | | 45. A○ | B○ | C○ | D○ | E○ |
| 16. A○ | B○ | C○ | D○ | E○ | | 46. A○ | B○ | C○ | D○ | E○ |
| 17. A○ | B○ | C○ | D○ | E○ | | 47. A○ | B○ | C○ | D○ | E○ |
| 18. A○ | B○ | C○ | D○ | E○ | | 48. A○ | B○ | C○ | D○ | E○ |
| 19. A○ | B○ | C○ | D○ | E○ | | 49. A○ | B○ | C○ | D○ | E○ |
| 20. A○ | B○ | C○ | D○ | E○ | | 50. A○ | B○ | C○ | D○ | E○ |
| 21. A○ | B○ | C○ | D○ | E○ | | 51. A○ | B○ | C○ | D○ | E○ |
| 22. A○ | B○ | C○ | D○ | E○ | | 52. A○ | B○ | C○ | D○ | E○ |
| 23. A○ | B○ | C○ | D○ | E○ | | 53. A○ | B○ | C○ | D○ | E○ |
| 24. A○ | B○ | C○ | D○ | E○ | | 54. A○ | B○ | C○ | D○ | E○ |
| 25. A○ | B○ | C○ | D○ | E○ | | 55. A○ | B○ | C○ | D○ | E○ |
| 26. A○ | B○ | C○ | D○ | E○ | | 56. A○ | B○ | C○ | D○ | E○ |
| 27. A○ | B○ | C○ | D○ | E○ | | 57. A○ | B○ | C○ | D○ | E○ |
| 28. A○ | B○ | C○ | D○ | E○ | | 58. A○ | B○ | C○ | D○ | E○ |
| 29. A○ | B○ | C○ | D○ | E○ | | 59. A○ | B○ | C○ | D○ | E○ |
| 30. A○ | B○ | C○ | D○ | E○ | | 60. A○ | B○ | C○ | D○ | E○ |

Name _____

**Mark the letter of the correct choice for the items shown in Practice Test 2.**

| | | | | | | | | | | | |
|---|---|---|---|---|---|---|---|---|---|---|---|
| 1. | A○ | B○ | C○ | D○ | E○ | 31. | A○ | B○ | C○ | D○ | E○ |
| 2. | A○ | B○ | C○ | D○ | E○ | 32. | A○ | B○ | C○ | D○ | E○ |
| 3. | A○ | B○ | C○ | D○ | E○ | 33. | A○ | B○ | C○ | D○ | E○ |
| 4. | A○ | B○ | C○ | D○ | E○ | 34. | A○ | B○ | C○ | D○ | E○ |
| 5. | A○ | B○ | C○ | D○ | E○ | 35. | A○ | B○ | C○ | D○ | E○ |
| 6. | A○ | B○ | C○ | D○ | E○ | 36. | A○ | B○ | C○ | D○ | E○ |
| 7. | A○ | B○ | C○ | D○ | E○ | 37. | A○ | B○ | C○ | D○ | E○ |
| 8. | A○ | B○ | C○ | D○ | E○ | 38. | A○ | B○ | C○ | D○ | E○ |
| 9. | A○ | B○ | C○ | D○ | E○ | 39. | A○ | B○ | C○ | D○ | E○ |
| 10. | A○ | B○ | C○ | D○ | E○ | 40. | A○ | B○ | C○ | D○ | E○ |
| 11. | A○ | B○ | C○ | D○ | E○ | 41. | A○ | B○ | C○ | D○ | E○ |
| 12. | A○ | B○ | C○ | D○ | E○ | 42. | A○ | B○ | C○ | D○ | E○ |
| 13. | A○ | B○ | C○ | D○ | E○ | 43. | A○ | B○ | C○ | D○ | E○ |
| 14. | A○ | B○ | C○ | D○ | E○ | 44. | A○ | B○ | C○ | D○ | E○ |
| 15. | A○ | B○ | C○ | D○ | E○ | 45. | A○ | B○ | C○ | D○ | E○ |
| 16. | A○ | B○ | C○ | D○ | E○ | 46. | A○ | B○ | C○ | D○ | E○ |
| 17. | A○ | B○ | C○ | D○ | E○ | 47. | A○ | B○ | C○ | D○ | E○ |
| 18. | A○ | B○ | C○ | D○ | E○ | 48. | A○ | B○ | C○ | D○ | E○ |
| 19. | A○ | B○ | C○ | D○ | E○ | 49. | A○ | B○ | C○ | D○ | E○ |
| 20. | A○ | B○ | C○ | D○ | E○ | 50. | A○ | B○ | C○ | D○ | E○ |
| 21. | A○ | B○ | C○ | D○ | E○ | 51. | A○ | B○ | C○ | D○ | E○ |
| 22. | A○ | B○ | C○ | D○ | E○ | 52. | A○ | B○ | C○ | D○ | E○ |
| 23. | A○ | B○ | C○ | D○ | E○ | 53. | A○ | B○ | C○ | D○ | E○ |
| 24. | A○ | B○ | C○ | D○ | E○ | 54. | A○ | B○ | C○ | D○ | E○ |
| 25. | A○ | B○ | C○ | D○ | E○ | 55. | A○ | B○ | C○ | D○ | E○ |
| 26. | A○ | B○ | C○ | D○ | E○ | 56. | A○ | B○ | C○ | D○ | E○ |
| 27. | A○ | B○ | C○ | D○ | E○ | 57. | A○ | B○ | C○ | D○ | E○ |
| 28. | A○ | B○ | C○ | D○ | E○ | 58. | A○ | B○ | C○ | D○ | E○ |
| 29. | A○ | B○ | C○ | D○ | E○ | 59. | A○ | B○ | C○ | D○ | E○ |
| 30. | A○ | B○ | C○ | D○ | E○ | 60. | A○ | B○ | C○ | D○ | E○ |

$$y = mx + b \quad m$$

$$y = mx + 5 - b$$

$$\frac{y - b}{nx} = \frac{mx}{x}$$

$$\frac{y - b}{x}$$

$$Km + 5x = 64$$

$$Km + 5x - 5x = 64 - 5x$$

$$\frac{Km}{K} = \frac{64 - 5x}{K}$$

$$m = \frac{64 - 5x}{K}$$

$$A = \frac{1}{2}bh \cdot 2$$

$$\frac{2a}{b} = \frac{bh}{b}$$

$$\frac{2a}{b} = h$$